QUICK ARITHMETIC

QUICK ARITHMETIC

ROBERT A. CARMAN
Santa Barbara City College
Santa Barbara, California

MARILYN J. CARMAN
Santa Barbara City Schools
Santa Barbara, California

John Wiley & Sons, Inc.
New York • London • Sydney • Toronto

Editors: Judy Wilson and Irene Brownstone
Production Manager: Ken Burke
Editorial Supervisor: Lorna Cunkle
Artist: Martha Hairston

Library of Congress Cataloging in Publication Data

Carman, Robert A
  Quick arithmetic

  (Wiley self-teaching guides)
   1. Arithmetic--1961-   2. Arithmetic--Programmed instruction.   I. Carman, Marilyn J., joint author.
II. Title
[QA107.058]   513'.07'7   74-2476
ISBN 0-471-13496-1

Printed in the United States of America

74 75 10 9 8 7 6 5 4 3 2 1

To

Bessie Sparrow

whose life

is an inspiration

to all who know her

Preface

Many students enrolled in our colleges, universities, and community colleges find themselves frustrated. They are eager, ambitious, and often quite capable of succeeding in their planned careers. They want to learn, but they find themselves handicapped because they do not have the basic mathematics skills needed to continue. They are not interested in "new math"; they need help with basic arithmetic computation skills— very old math indeed, but very used and very vital. If that describes your need, this book is for you.

This book is designed to help you review or relearn basic arithmetic skills. It is more like a private tutor than a lecturer; you participate in the process rather than simply read, listen, or sleep through it. The book is organized in a format that respects your unique needs and interests and teaches you accordingly. It may be used for self-study, for study with a tutor, or as a text in a formal course. Each chapter begins with a diagnostic preview or pretest and specific objectives keyed to the text, all designed to help you determine your particular needs. You have the option of skipping familiar material to save time or working through all of it if you must. Many practice problems and self-tests are included. At each step the programmed format provides feedback to assure your understanding. Unlike previous mathematics textbooks you may have used, this book is careful to explain every operation. Sometimes we even explain our explanations! This material has been tested on thousands of students and they tell us it is helpful, interesting, and even fun to work through. We hope you agree with them.

It is a pleasure for us to acknowledge our debts to the many people who have contributed to the development of this book. Irene Brownstone of John Wiley & Sons made many very helpful suggestions and the book has profited greatly from her efforts. Over the last few years, students and teachers at Santa Barbara City College have had the good fortune to work with a group of skilled and dedicated tutors. Their insights gained as tutors are a valuable part of this book. Especially helpful were Andy Aull, Tim Hall, Lynne Brown, Irma Herrera, Craig Turek, and David Castro. They taught us to listen and to care. We wish to extend special thanks to our daughter Patty, whose sharp wit, gentle criticism, and artistic ability made this a much better book than it would have otherwise been. Dr. Julio Bortolazzo deserves special thanks for his encouragement and guidance. His insistence on the dignity of all work and the dignity of all men forced us to explore this avenue of teaching. We are grateful.

Santa Barbara, California
September, 1974

Robert A. Carman
Marilyn J. Carman

Contents

How to Use This Book

Sally is one of the many people who go through life afraid of mathematics and upset by numbers. She will bumble along miscounting her marbles, bouncing checks, and eventually trying to avoid college courses that require even simple math. Most such people need to return and make a fresh start. Few get the chance. This book presents fresh-start math. It is designed so that you can:

- start at the beginning or where you need to start,
- work on only what you need to know,
- move as fast or as slowly as you wish,
- skip material you already understand,
- do as many practice problems as you need,
- take self-tests to measure your progress.

In other words, if you find mathematics difficult and want a fresh start, this book is designed for you.

This is no ordinary book. You cannot browse in it; you don't read it. You work your way through it. The ideas are arranged step-by-step in short portions or frames. Each frame contains information, careful explanations, examples, and questions to test your understanding. Read the material in each frame carefully, follow the examples, and answer the questions that lead to the next frame. Correct answers move you quickly through the book. Incorrect answers lead you to frames that provide further explanation. You move through this book frame by frame, sometimes forward, sometimes backward. Because we know that every person is different and has different needs, each major section of the book starts with a preview that will help you determine those parts on which you need to work.

As you move through the book you will notice that material not directly connected to the frames appears in boxes and margins. Read these at your leisure. They contain information that you may find useful and interesting.

Most students hesitate to ask questions. They would rather risk failure than look foolish by asking "dumb" questions. To relieve you of worry over dumb questions (or DAQs), we'll ask and answer them for you. Thousands of students have taught us that "dumb" questions can produce smart students. Watch for DAQ.

In 1846, the Reverend H. W. Adams described what happened when the ten-year-old math whiz Truman Safford was asked to multiply, in his head, the number 365,365,365,365,365,365 by itself. "He flew around the room like a top, pulled his pantaloons over the tops of his boots, bit his hands, rolled his eyes in their sockets, sometimes smiling and talking, and then, seeming to be in agony, in not more than one minute, he said 133,491,850,208,566,925,016,658,299,941,583,255." [*] In this book we will show you a way to do arithmetic that is not so strenuous, quite a bit slower, and not nearly so much fun to watch.

Now, turn to page 1 and let's begin.

---

[*] James R. Newman, The World of Mathematics (New York: Simon and Schuster, 1956), p. 466.

QUICK ARITHMETIC

CHAPTER ONE
Arithmetic of Whole Numbers

Preview

Objectives	Where to Go for Help	

Upon successful completion of this chapter you will be able to:

	Page	Frame

1. Add, subtract, multiply, and divide whole numbers.

	Page	Frame
(a) $6341 + 14,207 + 635 =$ _____	3	**1**
(b) $64,508 - 37,629 =$ _____	20	**16**
(c) $4328 \times 407 =$ _____	29	**24**
(d) $672 \times 2009 =$ _____	29	**24**
(e) $46,986 \div 745 =$ _____	41	**35**
(f) $37\overline{)3003} =$ _____	41	**35**
(g) $\dfrac{1541}{23} =$ _____	41	**35**
(h) $12 \times 0 =$ _____	29	**24**
(i) $16 \div 1 =$ _____	41	**35**

2. Write a whole number as a product of its prime factors.

	Page	Frame
(a) $3780 =$ _____	51	**44**
(b) $1848 =$ _____	51	**44**

		Page	Frame
3.	Calculate integer powers of a whole number.		
	(a) $2^3 =$ _____	65	**59**
	(b) $42^2 =$ _____	65	**59**
4.	Find the square root of a perfect square.		
	(a) $\sqrt{169} =$ _____	65	**59**

If you are certain you can work all of these problems correctly, turn to page 75 for a self-test. If you want help with any of these objectives or if you cannot work one of the preview problems, turn to the page indicated. Super-students (those who want to be certain they learn all of this), turn to frame **1** and begin work there.

ANSWERS TO PREVIEW PROBLEMS

4. (a) 13

3. (b) 1764
   (a) 8

2. (b) $2^3 \cdot 3 \cdot 7 \cdot 11$
   (a) $2^2 \cdot 3^3 \cdot 5 \cdot 7$

   (i) 16
   (h) 0
   (g) 67
   (f) 81 with remainder 6
   (e) 63 with remainder 51
   (d) 1,350,048
   (c) 1,761,496
   (b) 26,879
1. (a) 21,183

Arithmetic of Whole Numbers

1

© 1970 United Feature Syndicate, Inc.

Charlie Brown and his friends use numbers to tell time, count marbles, and keep track of their lunch money. Even for these simple operations "tooty-two" won't do the job. We may be living in an age of electronic computers and calculating machines, but most of the arithmetic used in industry, business, and school is still done by hand. For most educated adults, working with numbers is as important a part of their job as being able to read or write. In this chapter we will take a how-to-do-it look at the basic operations of arithmetic: addition, subtraction, multiplication, and division.

What is a number? It is a way of thinking, an idea, that enables us to compare very different sets of objects. It is the idea behind the act of counting. The number three is the idea that describes any collection of three objects: 3 people, 3 trees, 3 colors, 3 dreams. We recognize that these collections all have the quality of "threeness" even though they may differ in every other way.

We use numerals to name numbers. For example, the number of corners on a square is four, or 4, or IV in Roman numerals, or 囚 in Chinese numerals, or "tooty-two" for Sally in the cartoon.

In our modern number system we use ten digits—0, 1, 2, 3, 4, 5, 6, 7, 8, and 9—to build numerals just as we use the twenty-six letters of the alphabet to build words.

Is 10 a digit? Think about it. Then turn to **3** to continue.

2

Hi. What are you doing here? Lost? Window shopping? Just passing through? Nowhere in this book are you directed to frame **2**. (Notice that little **2** to the left above? That's a frame number.) Remember, in this book you move from frame to frame as directed, but not necessarily in 1-2-3 order. Follow directions and you'll never get lost.

Now return to **1** and keep working.

3

No, 10 is a numeral formed from the two digits 1 and 0. Remember:

- A <u>number</u> is an idea related to counting.

- A <u>numeral</u> is a symbol used to represent a number.

- A <u>digit</u> is one of the ten symbols (0, 1, 2, 3, 4, 5, 6, 7, 8, 9) we use to form numerals.

How many letters are in this set?

Count them. Write your answer, then turn to frame **4**.

4

We counted 23, and of course we write it in the ordinary, everyday manner. Leave the Roman, Chinese, and other numeral systems to Romans, Chinese, and people who enjoy the history of mathematics.

The basis of our system of numeration is grouping into sets of ten or multiples of ten.

    becomes

or   **A A A A A A A A A A**
      A A A A A A A A A A   **A A A**

2 tens + 3 ones
20   + 3   (or 23)

Numbers written as multiples of ten are said to be in <u>expanded</u> <u>form</u>. Any number may be written in this way. For example,

$$46 = 40 + 6 = 4 \text{ tens} + 6 \text{ ones}$$
$$274 = 200 + 70 + 4 = 2 \text{ hundreds} + 7 \text{ tens} + 4 \text{ ones}$$
$$305 = 300 + 0 + 5 = 3 \text{ hundreds} + 0 \text{ tens} + 5 \text{ ones}$$

Write out the following in expanded form:

362 = ____ + ____ + ____ = ____ hundreds + ____ tens + ____ ones

425 = ____ + ____ + ____ = ____ hundreds + ____ tens + ____ ones

208 = ____ + ____ + ____ = ____ hundreds + ____ tens + ____ ones

Check your work in **5.**

NAMING LARGE NUMBERS

Any large number given in numerical form may be translated to words by using the following diagram.

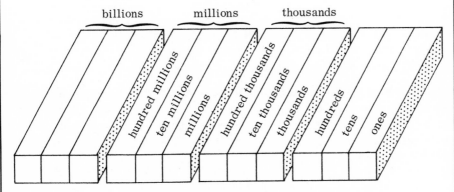

The number 14,237 can be placed in this diagram like this

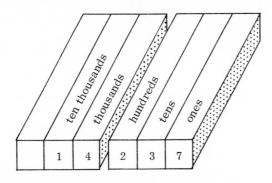

and is read "fourteen thousand, two hundred thirty-seven."

The number 47,653,290,866 becomes

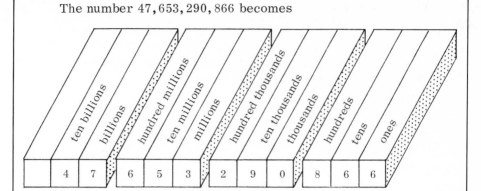

and is read "forty-seven billion, six hundred fifty-three million, two hundred ninety thousand, eight hundred sixty-six."

In each block of three digits read the digits in the normal way ("forty-seven," "six hundred fifty-three") and add the name of the block ("billion," "million"). Notice that the word "and" is not used in naming these numbers.

5

$362 = 300 + 60 + 2 = 3$ hundreds $+ 6$ tens $+ 2$ ones
$425 = 400 + 20 + 5 = 4$ hundreds $+ 2$ tens $+ 5$ ones
$208 = 200 + 0 + 8 = 2$ hundreds $+ 0$ tens $+ 8$ ones

Notice that the 2 in 362 means something very different from the 2 in 425 or 208. In 362 the 2 signifies two ones. In 425 the 2 signifies two tens. In 208 the 2 signifies two hundreds. Ours is a place value system of naming numbers: the value of any digit depends on the place where it is located.

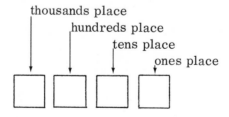

Writing numbers in expanded form will be helpful later when you want to understand how arithmetic operations work.

Addition

Addition is the simplest arithmetic operation.

$$4 + 3 = \underline{\hphantom{XXXX}}$$

Complete the calculation and go to **7**.

6

Here is the completed addition square:

Add	4	2	8	7	5	6	1	3	9
2	6	4	10	9	7	8	3	5	11
4	8	6	12	11	9	10	5	7	13
7	11	9	15	14	12	13	8	10	16
5	9	7	13	12	10	11	6	8	14
1	5	3	9	8	6	7	2	4	10
9	13	11	17	16	14	15	10	12	18
6	10	8	14	13	11	12	7	9	15
8	12	10	16	15	13	14	9	11	17
3	7	5	11	10	8	9	4	6	12

Did you notice that changing the order in which you add numbers does not change their sum?

$$4 + 3 = 7 \text{ and } 3 + 4 = 7$$
$$2 + 4 = 6 \text{ and } 4 + 2 = 6$$

This is true for any addition problem involving whole numbers. It is known as the commutative property of addition. (A commuter is a person who changes location daily, moving back and forth between suburbs and city. The commutative property says changing the location or order of the numbers being added does not change their sum.)

If you have not already memorized the addition of one-digit numbers, it is time to do so.   To help you, a study card for addition facts is provided in the back of this book on page 239.   Use it if you need it.

If you want more practice adding one-digit numbers, go to **9**.   Otherwise, continue in **8**.

7

$$4 \quad + \quad 3 \quad = \quad 7$$

We add collections of objects by combining them into a single set and then counting and naming that new set.   The numbers being added (4 and 3 in this case) are called <u>addends</u> and 7 is the <u>sum</u> of the addition.

There are a few simple addition facts you should have stored in your memory and ready to use.   Complete the following table by adding the number at the top to the number at the side and placing their sum in the proper square.   We have added 1 + 2 = 3 and 4 + 3 = 7 for you.

Add	4	2	8	7	5	6	1	3	9
2									
4								7	
7									
5									
1		3							
9									
6									
8									
3									

Check your answer in **6.**

ROMAN NUMERALS

A number is an idea. A numeral is a symbol that enables us to express that idea in writing and use it in counting and calculating. Roman numerals were used by the ancient Romans almost 2000 years ago and are still seen on clock faces, building inscriptions, and textbooks. The following seven symbols are used:

I	V	X	L	C	D	M
1	5	10	50	100	500	1000

Notice that the numbers represented are 1, 5, and multiples of 5 and 10 (the number of fingers on one hand and on two hands). There is no zero. We write numerals with these symbols by placing them in a row and adding or subtracting. For example,

1 = I	7 = VII (V + I + I)
2 = II	8 = VIII
3 = III	9 = IX (I subtracted from X)
4 = IV (I subtracted from V)	10 = X
5 = V	27 = XXVII
6 = VI (V + I)	152 = CLII

The Romans used only addition and they wrote 4 as IIII, but in order to keep numerals smaller, later mathematicians used subtraction to form numbers like these:

IV	IX	XL	XC	CD	CM
4	9	40	90	400	900

Only these six subtractions are allowed. From these other combinations can be made.

XIX = X + IX or 10 + 9 or 19

LXIV = LX + IV or 60 + 4 or 64

Roman numerals are a bit more difficult to write than the ones we use and they are a headache to multiply or divide, but they are very easy to add or subtract. For example, 111 + 16 = 127 would be written like this:

CXI + XVI = CXXVII

The numerals we use now (0, 1, 2, 3, etc.) were first seen in Europe in about the thirteenth century but Roman numerals were used by bankers and bookkeepers until the eighteenth century. They did not trust symbols like 0 that could easily be changed to 6, 8, or 9 by a dishonest clerk.

8

Now, let's try a more difficult addition problem.

$$35 + 42 = \underline{\hspace{1cm}}$$

The first step in any arithmetic problem is to <u>guess</u> at the answer. Never work a problem until you know roughly what the answer is going to be. Always know where you are going.

Make a guess at the answer to the problem above. Write your guess here _____ and then turn to **10** and continue.

I called 4+2 "four and two" and my math teacher turned green. Why?

Call 4+2 "four plus two." There is no math operation called "and." Your instructor gets upset because he's not sure what you are saying and he's afraid you don't know either.

9

Problem Set 1: Practice Problems for One-Digit Addition

Add the following. Work quickly. You should be able to answer all problems in a set correctly in the time indicated. (The times are for community college students enrolled in a developmental math course.) Try to do all addition mentally.

A. Add

3	7	3	8	5	3	9	2	6	8
5	9	3	5	6	8	4	7	7	4

7	9	4	7	8	6	7	9	8	5
7	8	2	5	7	3	4	3	8	4

9	2	6	2	8	4	2	5	9	8
6	8	4	9	6	3	6	9	9	7

8	5	7	8	9	6	9	7	8	6
6	9	5	4	7	6	4	6	5	9

7	8	6	3	8	7	6	4	9	5
4	9	5	9	3	8	7	8	9	7

Average Time: 90 seconds; Record: 32 seconds

B. Add

7	5	2	5	8	4	3	8	9	7
3	6	9	7	8	5	6	4	3	6

6	8	9	3	7	2	9	9	7	4
4	5	6	5	7	7	4	9	8	7

8	9	8	5	8	4	9	6	4	8
3	7	6	5	9	5	5	6	3	2

5	6	7	7	5	6	2	3	6	9
8	7	5	9	4	5	8	7	8	8

7	5	9	4	3	8	7	8	5	7
4	9	2	6	8	6	9	4	8	8

Average Time: 90 seconds; Record: 35 seconds

C. Add

2	7	3	4	2	6	3	5	9	5
5	3	6	5	7	7	4	7	6	2
4	2	5	8	9	8	4	8	3	8

6	5	4	8	6	9	7	4	8	1
2	4	2	1	8	3	1	9	4	8
7	5	9	9	8	5	6	1	6	7

1	9	3	1	7	2	9	9	8	5
9	9	1	6	9	9	8	5	3	4
2	1	4	3	6	1	2	1	3	7

Average Time: 90 seconds; Record: 41 seconds

The answers to the problems on pages 10 and 11 are on page 222. When you have had the practice you need, turn to **8** and continue.

10

35 + 42 is approximately 30 + 40 or 70. The correct answer will be about 70, not 7 or 700 or 7000. Once you have a reasonable estimate of the answer, you are ready to do the arithmetic work.

$$35 + 42 = \underline{\hspace{1cm}}$$

Check your answer in **11**.

11

You should have set the problem up like this:

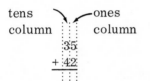

tens   ones
column   column

▶1. The numbers to be added are arranged vertically in columns.
▶2. The right end or ones digits are placed in the ones column, the tens digits are placed in the tens column, and so on.

The most frequent cause of errors in arithmetic is carelessness, especially in simple procedures such as lining up the digits correctly.

Avoid the confusion of

$$\begin{array}{r} 35 \\ + 42 \\ \hline \end{array} \quad \text{or} \quad \begin{array}{r} 35 \\ + 42 \\ \hline \end{array}$$

Once the digits are lined up the problem is easy.

$$\begin{array}{r} 35 \\ + 42 \\ \hline 77 \end{array}$$

Does your answer agree with your original guess? Yes. The guess, 70, is roughly equal to the actual sum, 77.

What we have just shown you is the guess 'n check method of doing mathematics calculations.

- ■ Step 1: Guess at the answer.
- ■ Step 2: Work the problem carefully.
- ■ Step 3: Check your answer against your guess. If they are very different, repeat both step 1 and 2.

Most students hesitate to guess, afraid they might guess incorrectly. Relax. You are the only one who will know your guess. Do it in your head, do it quickly, and make it reasonably accurate. Step 3 helps you detect incorrect answers before you finish the problem. The guess 'n check method means you never work in the dark, you always know where you are going. Use this approach on every math calculation and you need never have an incorrect answer again.

Here is a slightly more difficult problem. Try it, then go to **12**.

$27 + 48 = $ _____

I still count on my fingers when I add. Is that bad?

Not really. It's slow and sometimes inconvenient, but it is the normal way to start. Keep working to memorize one-digit number additions.

12

First, guess.   27 + 48 is roughly 30 + 50 or about 80. The answer is closer to 80 than to 8 or 800.

Second, line up the addends vertically.   27
                                                 + 48

Third, work it out carefully.   ¹
                                       27
                                      + 48
                                       75

Finally, check your answer   The guess, 80, is roughly equal against your guess.   to the actual answer, 75.

What does that little **/** above the tens column mean? What really happens when you "carry" a digit? In expanded notation:

    27 = 2 tens + 7 ones
  + 48 = 4 tens + 8 ones
         6 tens + 15 ones = 6 tens + 1 ten + 5 ones
                          = 7 tens + 5 ones
                          = 75

The **/** that is "carried" over to the tens column is really a 10! Another way to see this is shown on the next page.

■Step 1   ■Step 2   ■Step 3

27   27   27
+ 48 add ones   + 48   + 48
15   7 + 8 = 15   15   add tens   15
                          60   20 + 40 = 60   60 add partial sums
                                                     75 15 + 60 = 75

15 is the first   60 is the second
partial sum.   partial sum.   75 is the final sum.

Again, you should see that the "carry 1" is the 10 in 15.

Using partial sums is the long way to add, so usually we take a short-cut and write:

      l   *l*

     27   7 + 8 = 15   27

   + 48   Write 5,   + 48

      5   carry *l* ten.   75   *l* + 2 + 4 = 7

You will learn the short-cut method here, but it is important that you know why it works.

Use the partial sum method to work this problem.

    429 + 758 = _____

Set the problem up step by step as we did above, then turn to **13**. (Don't forget to guess 'n check.)

13

    429 + 758 = _____

Guess: 400 + 700 = 1100

Line up the addends: 429
                       + 758

Use partial sums:

       429
     + 758
       17   add ones: 9 + 8 = 17
       70   add tens: 20 + 50 = 70
   1100   add hundreds: 400 + 700 = 1100
   1187   add the partial sums

Check: 1100 (the guess) roughly equals 1187.

And of course you would use the short-cut method (shown on the next page) once you understand this process.

■Step 1	■Step 2	■Step 3

$\overset{1}{4}29$ 9 + 8 = 17

+ 758   Write 7,

7   carry 1 ten.

$\overset{1}{4}29$

+ 758   1 + 2 + 5 = 8

87

$\overset{1}{4}29$

+ 758   4 + 7 = 11

1187

Now try these problems. Work them first using partial sums, then using the short-cut. Be sure to guess 'n check.

(a) 256 + 867 = _____

(b) 2368 + 754 = _____

(c) 980 + 456 = _____

Check your answers in **14** when you are finished.

HOW TO ADD LONG LISTS OF NUMBERS

Very often, especially in business and industry, it is necessary to add long lists of numbers. The best procedure is to break the problem down into a series of simpler additions. First add sets of two or three numbers, then add these sums to obtain the total. For example,

```
      9
      3        12
      7
      6     13      25
     12
      4        16
     17
   +  5     22      38
                    63
```

Using this procedure you do a little more writing but carry fewer numbers in your head. The result is fewer mistakes.

Better yet, keep your eye open for combinations that add to 10 or 15, and work with mental addition of three addends.

```
      9
      3 ⎫
      7 ⎬— 10      19
      6 ⎫
     12 ⎬— 10
      4 ⎭          22
     17
   +  5            22
                   63
```

Try these for practice:

8	7	3	11	3	13
17	6	5	7	5	17
3	8	7	2	12	11
4	5	6	5	7	14
11	9	5	6	6	15
9	3	1	7	4	8
16	7	3	13	1	9
7	12	4	6	2	16
11	8	2	5	18	12
5	16	7	14	9	7
		3	16	7	18

The answers are on page 222.

14

(a) Guess: 200 + 900 = 1100

```
   256
 + 867
    13   add ones:  6 + 7 = 13
   110   add tens:  50 + 60 = 110
  1000   add hundreds:  200 + 800 = 1000
  1123   add partial sums
```

Check: 1100 is roughly equal to 1123.

Short-cut method:

■Step 1	■Step 2	■Step 3
256 6 + 7 = 13	256 1 + 5 + 6 = 12	256 1 + 2 + 8 =
+ 867 Write 3,	+ 867 Write 2,	+ 867 11
3 carry 1 ten.	23 carry 1 hundred.	1123

(b) Guess: 2300 + 700 = 3000

```
   2368
 +  754
     12   add ones:  8 + 4 = 12
    110   add tens:  60 + 50 = 110
   1000   add hundreds:  300 + 700 = 1000
   2000   add thousands:  2000
   3122   add partial sums
```

Check: The guess, 3000, is roughly equal to 3122.

(b) continued

Short-cut method:

■Step 1

2368   8 + 4 = 12
+ 754   Write 2,
2   carry 1 ten.

■Step 2

2368   1 + 6 + 5 = 12
+ 754   Write 2,
22   carry 1 hundred.

■Step 3

2368   1 + 3 + 7 = 11
+ 754   Write 1,
122   carry 1 thousand

■Step 4

2368   1 + 2 = 3
+ 754
3122

(c) Guess: 1000 + 400 = 1400

$$
\begin{array}{r}
980 \\
+ 456 \\
\hline
6 \\
130 \\
1300 \\
\hline
1436
\end{array}
$$

add ones: 0 + 6 = 6
add tens: 80 + 50 = 130
add hundreds: 900 + 400 = 1300

Check: 1400 is approximately equal to 1436.

Short-cut method:

■Step 1

980   0 + 6 = 6
+ 456
6

■Step 2

980   8 + 5 = 13
+ 456   Write 3,
36   carry 1 hundred.

■Step 3

980   1 + 9 + 4 = 14
+ 456
1436

If you had difficulty with any of these problems, you should return to
5 and review. Otherwise go to **15** for a set of practice addition problems.

15

Problem Set 2: Addition

A. Add

| 47 | 18 | 27 | 57 | 45 | 89 |
| 23 | 86 | 38 | 69 | 35 | 17 |

| 73 | 44 | 92 | 38 | 88 | 75 |
| 39 | 28 | 39 | 65 | 17 | 48 |

47	26	76	48	33	67
56	98	24	84	19	69

B. Add

273	189	726	508	701
142	204	387	495	829

684	729	432	708	621
706	287	399	554	388

386	747	593	375	906
438	59	648	486	95

C. Add

4237	6489	5076	1684
1288	3074	4385	927

7907	1467	3015	9864
1395	2046	687	2735

6872	8360	6009	3785
493	1762	496	7643

5049	6709	8475	6008
732	9006	928	5842

D. Add

18745	10674	60485	12008
6972	397	9766	9634

9876	59684	40026
4835	29527	7085

78044	94036	87468
97684	6975	92729

E. Arrange vertically and add

487 + 29 + 526 =

715 + 4293 + 184 + 19 =

1706 + 387 + 42 + 307 =

456 + 978 + 1423 + 3584 =

6284 + 28 + 674 + 97 =

6842 + 9008 + 57 + 368 =

322 + 46 + 5984 =

7268 + 209 + 178 =

5016 + 423 + 1075 =

8764 + 85 + 983 + 19 =

F. Brain Boosters (Brain Boosters are more difficult and more fun than the regular problems. You will find them challenging, but don't expect to be as successful with them as you are with the others.)

1. The Clinker University tiddley wink team weighs in as follows: "Tank" Murphy, 263 lb; Bertha Brown, 218 lb; "Moose" Green, 314 lb; and Head Tiddler, "Tiger" Smith, 87 lb. What is the total weight of the Clinker Winkers?

2. At lunch the other day, Mary the calorie counter ate the following: one slice of whole wheat bread, 55 calories; cream cheese and honey on the bread, 148 calories; yogurt, 123 calories; fresh blackberries, 45 calories. What was her total calorie count for the meal?

3. Texas has more miles of highway than any other state. In 1970 Texas had a road mileage of 237,769; California was next with 162,809 miles; Kansas third with 133,232 miles; and Illinois fourth with 128,479 miles. What was the total road mileage for these four states?

4. During the first three months of the year, the Write Rite Typewriter Company reported the following sales:

January	$3572
February	$2716
March	$4247

What was their sales total for this quarter of the year?

5. Which sum is greater?

987654321	123456789
87654321	123456780
7654321	123456700
654321	123456000
54321	123450000
4321	123400000
321	123000000
21	120000000
1	100000000

6. In the Munich Olympic games in 1972 Nikolay Avilov of the Soviet Union won the decathlon with a world record point total. In the ten events he scored as follows:

100 meter run	804		High hurdles	926
Long jump	957		Discus	818
Shot put	750		Pole vault	945
High jump	959		Javelin	781
400 meter run	875		1500 meter run	639

What was his total score?

The answers to these problems are on pages 222-223. When you have completed these practice problems you may continue in **16** with the study of subtraction or return to the preview for this chapter on page 1 and use it to determine the help you need next.

Subtraction

16

Subtraction is the reverse of the process of addition.

Addition: $3 + 4 = \Box$

Subtraction: $3 + \Box = 7$

Written this way, a subtraction problem asks the question, "How much must be added to a given number to produce a required amount?" Most often, however, the numbers in a subtraction problem are written using a minus sign (-):

$17 - 8 = \Box$

This means that there is a number $\Box$ such that $8 + \Box = 17$. Write in the answer to this subtraction problem, then turn to **17.**

17

$$17 \quad - \quad 8 \quad = \quad 9$$

minuend   subtrahend   difference

Special names are given to the numbers in a subtraction problem and it will be helpful if you know them.

- The <u>minuend</u> is the larger of the two numbers in the problem. It is the number that is being decreased.

- The <u>subtrahend</u> is the number that is being subtracted from the minuend.

- The <u>difference</u> is the amount that must be added to the subtrahend to produce the minuend. It is the answer to the subtraction problem.

The ability to solve simple subtraction problems depends on your knowledge of the addition of one-digit numbers. For example, find this difference:

$9 - 4 =$ _____

Do this problem and then turn to **18**.

18

$9 - 4 = 5$

Solving this problem probably involved a chain of thoughts something like this:

"Nine minus four. Four added to what number gives nine? Five? Try it: four plus five equals nine. Right."

If you have memorized the addition of one-digit numbers (as shown in frame **6** or on the study card on page 239) subtraction problems involving small whole numbers will be easy for you. If you haven't memorized these, do it now.

Now try a more difficult subtraction problem.

$47 - 23 =$ _____

What is the first step? Work the problem and continue in **19**.

19

The <u>first</u> step is to guess at the answer! Remember?

47 – 23 is roughly 40 – 20 or 20.

The difference, your answer, will be about 20—not 2 or 10 or 200.

The <u>second</u> step is to write the numbers in a vertical format as you did with addition. Be careful to keep the ones digits in line in one column, the tens digits in a second column, and so on. Notice that the minuend is written above the subtrahend—larger number on top.

$$
\begin{array}{r}
47 \\
- 23 \\
\hline
\end{array}
$$

Once the numbers have been arranged this way the difference may be written immediately.

■Step 1   ■Step 2

$$
\begin{array}{r}
47 \\
- 23 \\
\hline
4
\end{array}
$$ ones digits: 7 – 3 = 4

$$
\begin{array}{r}
47 \\
- 23 \\
\hline
24
\end{array}
$$ tens digits: 40 – 20 = 20

The difference is 24 and this agrees with our first guess.

With some problems it is necessary to rewrite the minuend (larger number) before the problem can be solved. For example, the number 24 means 20 + 4 or 2 tens + 4 ones. It can be rewritten as 10 + 14. The difference 24 – 8 is not easy to find because 8 is larger than 4. To perform the subtraction we rewrite 24 as 10 + 14 and then subtract 8.

24 – 8 = (10 + 14) – 8 = 10 + (14 – 8) = 10 + 6 = 16

Try this one:

64 – 37 = _____

Check your work in **20.**

20

First, guess. 64 – 37 is roughly 60 – 40 or 20.

Second, arrange the numbers vertically in columns.

$$\begin{array}{r} 64 \\ -\ 37 \\ \hline \end{array}$$

Third, write them in expanded form to understand the process.

$$\begin{array}{rll} 64 = & 6 \text{ tens} + 4 \text{ ones} = & 5 \text{ tens} + 14 \text{ ones} \\ -\ 37 = & -\ (3 \text{ tens} + 7 \text{ ones}) = & -\ (3 \text{ tens} + 7 \text{ ones}) \\ \hline & = & 2 \text{ tens} + 7 \text{ ones} \\ & = & 20 + 7 \\ & = & 27 \text{ (which agrees with our guess)} \end{array}$$

Because 7 is larger than 4 we must "borrow" one ten from the six tens in the minuend. We are actually regrouping or rewriting the minuend.

In actual practice we do not write out subtraction problems in expanded form. Our work might look like this:

■Step 1   ■Step 2   ■Step 3

$$\begin{array}{r} 64 \\ -\ 37 \\ \hline \end{array}$$
$$\begin{array}{r} {}^{5}\ {}^{14} \\ \cancel{64} \\ -\ 37 \\ \hline 7 \end{array}$$ Borrow one ten, change the 6 in the tens place to **5**, change 4 to **14**, subtract **14** – 7 = 7.
$$\begin{array}{r} {}^{5}\ {}^{14} \\ \cancel{64} \\ -\ 37 \\ \hline 27 \end{array}$$ **5** 0 – 30 = 20, write 2.

Double check subtraction problems by adding the answer and the subtrahend. Their sum should equal the minuend.

$$\begin{array}{r} {}^{1} \\ 37 \\ +\ 27 \\ \hline 64 \end{array}$$

Try these problems for practice.

(a)   71   (b)   263   (c)   426   (d)   902
     - 39   - 127   - 128   - 465

Solutions are in **21**.

How do I subtract 90-25.4?

Easy. Write 90 as 90.0, then line up the decimal points and subtract as usual.
90.0
-25.4

MEASUREMENT NUMBERS

Many numbers are the result of measurement. A measurement number has two parts: a number part giving the size of the quantity and a unit giving a comparison standard for the measurement. For example, the winning time for a 100-yard dash is measured to be 9.6 seconds.

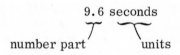

9.6 seconds

number part   units

Money numbers always have units: $5 or 25¢ are quantities measured in dollars and cents units.

To add or subtract measurement numbers:

▶Example:

1. Convert all numbers to the same units.

    2 feet + 10 inches =
    24 inches + 10 inches =

2. Add or subtract the number parts.

    24 + 10 = 34

3. Attach the common units to the sum or difference

    34 inches

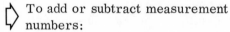

To multiply or divide measurement numbers:

▶Example:

1. Multiply or divide the number parts.

$$10 \text{ ft} \times 6 \text{ ft} =$$
$$10 \times 6 = 60$$

2. Attach the product or quotient of the units.

$$\text{ft} \times \text{ft} = \text{ft}^2$$
$$10 \text{ ft} \times 6 \text{ ft} = 60 \text{ ft}^2$$

	Common English Units	Common Metric Units
Length or Distance	inches, feet, miles	centimeters, meters, kilometers
Time	seconds, minutes, hours, days	seconds, minutes, hours, days
Weight or Mass	ounces, pounds	grams, kilograms
Volume	cup, quart, gallon	cubic centimeter, liter
Area	square inch, square foot, acre	centimeter squared (cm^2) meter squared (m^2)

Metric Equivalents

1 meter	a little larger than a yard, about 1.09 yd
1 centimeter	the width of a paper clip, about 0.4 inches
1 kilometer	about 0.6 miles
1 liter	a little larger than one quart, about 1.06 qt
1 cubic centimeter	about one fifth of a teaspoonful
1 kilogram	a little more than 2 pounds, about 2.2 lb

21

(a) Guess: 70 – 40 = 30

■Step 1   ■Step 2

    71   6 11
  – 39   7̸1̸
  ─────   – 39
                  ─────
                   32

Borrow one ten from 70,
change the 7 in the tens place to 6,
change the 1 in the ones place to 11,
11 – 9 = 2, write 2,
60 – 30 = 30, write 3.

Check: The answer 32 is approximately equal to the guess, 30.

(b) Guess: 200 - 100 = 100

■Step 1   ■Step 2

263        $\overset{5\ 13}{2\cancel{6}\cancel{3}}$

- 127   - 127

           136

Borrow one ten from 60,
change the 6 in the tens place to 5,
change the 3 in the ones place to 13,
13 - 7 = 6, write 6,
50 - 20 = 30, write 3,
200 - 100 = 100, write 1.

Check: The answer is approximately equal to the guess.

(c) Guess: 400 - 100 = 300

■Step 1   ■Step 2   ■Step 3

426        $\overset{1\ 16}{4\cancel{2}\cancel{6}}$ $\overset{3\ \ 11\ 16}{\cancel{4}\cancel{2}\cancel{6}}$

- 128   - 128   - 128

           8   298

16 - 8 = 8,
110 - 20 = 90, write 9,
300 - 100 = 200, write 2.

Notice that in this case we must borrow twice. Borrow one ten from
the 20 in 426 to make 16, then borrow one hundred from the 400 in
426 to make 110.

(d) Guess: 900 - 500 = 400

■Step 1   ■Step 2   ■Step 3

902        $\overset{8\ 10}{\cancel{9}\cancel{0}2}$ $\overset{8\ 9\ 12}{\cancel{9}\cancel{0}\cancel{2}}$

- 465   - 465   - 465

                              437

12 - 5 = 7,
90 - 60 = 30, write 3,
800 - 400 = 400, write 4.

In this problem we first borrow one hundred from 900 to get a 10 in
the tens place, then we borrow one ten from the tens place to get a
12 in the ones place. In expanded form problem (d) looks like this:

902   9 hundreds +   0 tens +   2 ones
- 465   - (4 hundreds +   6 tens +   5 ones)

                 8 hundreds + 10 tens +   2 ones
               - (4 hundreds +   6 tens +   5 ones)

                 8 hundreds +   9 tens + 12 ones
               - (4 hundreds +   6 tens +   5 ones)
                 4 hundreds +   3 tens +   7 ones

           =   400 + 30 + 7
           =   437

Do you want more worked examples of subtraction problems containing zeros, similar to this last one? If so, go to **23**. Otherwise, go to
22 for a set of practice problems.

22

Problem Set 3: Subtraction

A. Subtract

13	9	12	15	8	13	8	7	11	6
7	4	5	9	6	8	0	7	7	5

16	16	10	13	5	18	12	10	11	5
7	8	7	6	5	9	9	3	8	1

10	8	14	13	12	9	15	11	9	12
2	7	6	9	3	2	6	4	0	8

14	9	9	1	11	14	15	11	12	17
5	3	6	0	5	8	7	6	7	9

13	15	18	14	16	7	12	17	0	9
6	8	0	7	9	4	4	8	0	7

B. Subtract

40	63	78	33	51	85	36	60	42
27	19	49	17	39	28	17	43	27

91	52	47	70	94	34	55	56	93
63	16	29	48	57	9	29	18	8

C. Subtract

546	640	409	914	476	219	747	564
357	182	324	37	195	43	593	298

400	316	803	327	632	525	438	701
127	118	88	276	58	480	409	556

D. Subtract

6218	8704	6084	30209	13042	8000
3409	923	386	1367	524	321

| 57022 | 46804 | 5007 | 10785 | 10000 | 31072 |
| 980 | 9476 | 266 | 888 | 386 | 4265 |

| 48093 | 384000 | 27004 | 60754 | 42003 |
| 500 | 67360 | 4582 | 5295 | 17064 |

E.   Brain Boosters

1.   The enrollment at Sunshine Tech is 5804. If 3,985 students are women, how many male students are there?

2.   Eric had a bad day at marbles. He started the day with 461 and arrived home after school with 177. How many marbles did he lose?

3.   Last year on this date Sam's car odometer read 67,243 miles. It now reads 81,062 miles. How many miles has he driven in the past year?

4.   If your income is $8,245 per year and you pay $956 in taxes, what is your take-home pay?

5.   A $650 color TV is on sale for $495. How much money does Denny Doright save if he buys it at the sale price?

6.   Subtract nine thousand nine hundred nine from twelve thousand twelve hundred twelve.

7.   If you take three apples from a dish containing 13 apples, how many apples do you have?

8.   On March 1, Mrs. Pennywatcher had a balance of $635 in her checking account. During March she deposited checks of $352 and $114, and wrote checks for $37, $216, $147, and $106. How much did she have left in her account at the end of the month?

9.   Place "+" or "-" signs in each of the following sequences so that each one will total 100.

(a)   98 ___ 76 ___ 54 ___ 3 ___ 21 = 100

(b)   123 ___ 45 ___ 67 ___ 8 ___ 9 = 100

(c)   12 ___ 3 ___ 4 ___ 5 ___ 6 ___ 7 ___ 89 = 100

(d)   123 ___ 4 ___ 5 ___ 67 ___ 89 = 100

Can you make up more of these?

10.   Is it true that 1963 pennies are worth almost $20?

The answers to the problems on pages 27 and 28 are on page 223. When you have had the practice you need, go to **24** to study multiplication of whole numbers or return to the preview for this chapter on page 1.

23

Let's work through a few examples.

(a) ■Step 1   ■Step 2   ■Step 3   Check: 167
                  3 10   3 9 10   + 233
     400          $\cancel{4}\cancel{0}$0 $\cancel{4}\cancel{0}\cancel{0}$ 400
   - 167   - 167   - 167
                                233

Do you see that we have rewritten 400 as 300 + 90 + 10?

(b) ■Step 1   ■Step 2   ■Step 3   ■Step 4   Check:
                  4 10   4 9 10   4 99 16
     5006 $\cancel{5}$006 $\cancel{5}\cancel{0}$06 $\cancel{5}\cancel{0}\cancel{0}\cancel{6}$ 2487
   - 2487   - 2487   - 2487   - 2487   + 2519
                                                    2519   4006

(c) ■Step 1   ■Step 2   ■Step 3   ■Step 4   Check:
                   2 12   3 16 2 12   , 13 16 2 12
    24,632   246$\cancel{3}\cancel{2}$ 2$\cancel{4}\cancel{6}\cancel{3}\cancel{2}$ $\cancel{2}\cancel{4}\cancel{6}\cancel{3}\cancel{2}$ 5718
   - 5,718   - 5718   - 5718   - 5718   + 18914
                               14   914   18914   24632

Any subtraction problem that involves borrowing should <u>always</u> be checked. It is very easy to make a mistake in the borrowing process. Go to **22** for a set of practice problems on subtraction.

Multiplication

24

In a certain football game, the West Newton Waterbugs scored five touchdowns at six points each. How many total points did they score through touchdowns? We can answer the question several ways.

1. Count points: ·····························

2. Add touchdowns: 6 + 6 + 6 + 6 + 6 = ?

3. Multiply: 5 × 6 = ?

We're not sure about the mathematical ability of the West Newton scorekeeper, but most people would multiply. Multiplication is a short-cut method of counting or repeated addition.

How many points did the Waterbugs score? Work it out one way or another, then go to **25**.

25

$$5 \quad \times \quad 6 \quad = \quad 30$$

multiplier   multiplicand   product

In a multiplication statement the <u>multiplicand</u> is the number to be multiplied, the <u>multiplier</u> is the number multiplying the multiplicand, and the <u>product</u> is the result of the multiplication. The multiplier and multiplicand are called the <u>factors</u> of the product.

Notice that we can arrange these letters

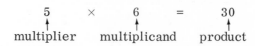

into 3 rows of 4

a a a a
a a a a
a a a a $3 \times 4 = 12$

or 4 rows of 3

a a a
a a a
a a a $4 \times 3 = 12$
a a a

Changing the order of the factors does not change their product. This is the <u>commutative</u> <u>property</u> <u>of</u> <u>multiplication</u>.

In order to become skillful at multiplication, you must know the one-digit multiplication table from memory.

Complete the table on the next page by multiplying the number at the top by the number at the side and placing their product in the proper square. We have multiplied $3 \times 4 = 12$ and $2 \times 5 = 10$ for you.

Multiply	2	5	8	1	3	6	9	7	4
1									
7									
5	10								
4					12				
9									
2									
6									
3									
8									

Check your answers in **27.**

26

Problem Set 4: Practice Problems for One-Digit Multiplication

Multiply as shown. Work quickly. You should be able to answer all problems in a set correctly in the time indicated. (These times are for community college students enrolled in a developmental math course.)

A. Multiply

6	4	9	6	3	9	7	8	2	8
2	8	7	6	4	2	0	3	7	1

6	8	5	5	2	3	9	7	3	0
8	2	9	6	5	3	8	5	6	4

7	5	4	7	4	8	6	9	8	6
4	3	9	7	2	5	7	6	8	4

5	3	5	9	9	6	1	8	4	7
4	0	5	3	9	1	1	6	4	9

Average Time: 100 seconds; Record: 37 seconds

B. Multiply

2	6	3	5	6	4	4	8	2	7
8	5	3	7	3	5	7	6	6	9

8	0	2	3	1	5	6	9	5	8
4	6	9	8	9	5	4	5	2	9

3	7	5	6	9	2	7	8	9	2
5	7	8	9	4	4	6	8	0	2

5	9	1	8	6	4	9	0	2	7
5	3	7	7	6	3	9	4	1	8

Average Time: 100 seconds; Record: 36 seconds

The answers to these problems are on page 223. When you have had the practice you need, turn to **28** and continue.

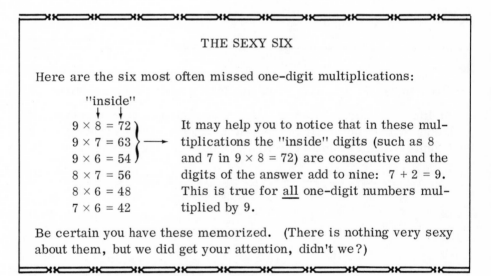

THE SEXY SIX

Here are the six most often missed one-digit multiplications:

"inside"

$9 \times 8 = 72$
$9 \times 7 = 63$ $\longrightarrow$ It may help you to notice that in these mul-
$9 \times 6 = 54$ tiplications the "inside" digits (such as 8
and 7 in $9 \times 8 = 72$) are consecutive and the
$8 \times 7 = 56$ digits of the answer add to nine: $7 + 2 = 9$.
$8 \times 6 = 48$ This is true for all one-digit numbers mul-
$7 \times 6 = 42$ tiplied by 9.

Be certain you have these memorized. (There is nothing very sexy about them, but we did get your attention, didn't we?)

27

The completed multiplication table is on the next page. If you are not able to perform these one-digit multiplications quickly from memory, you should practice until you can do so. A multiplication table is provided in the back of this book (page 241). Use it if you need it.

Multiply	2	5	8	1	3	6	9	7	4
1	2	5	8	1	3	6	9	7	4
7	14	35	56	7	21	42	63	49	28
5	10	25	40	5	15	30	45	35	20
4	8	20	32	4	12	24	36	28	16
9	18	45	72	9	27	54	81	63	36
2	4	10	16	2	6	12	18	14	8
6	12	30	48	6	18	36	54	42	24
3	6	15	24	3	9	18	27	21	12
8	16	40	64	8	24	48	72	56	32

Notice that the product of any number and one is that same number. For example,

$$1 \times 2 = 2$$
$$1 \times 6 = 6$$
$$1 \times 753 = 753$$

Zero has been omitted from the multiplication table because the product of any number and zero is zero. For example,

$$0 \times 2 = 0$$
$$0 \times 7 = 0$$
$$0 \times 395 = 0$$

If you want more practice in one-digit multiplication, go to **26.** Otherwise, go to **28.**

28

The multiplication of larger numbers is based on the one-digit number multiplication table. Find this product.

$$34 \times 2 = \underline{\hspace{2cm}}$$

Remember the procedure you followed for addition. What are the first few steps in this multiplication? Try it, then go to **29.**

29

First, guess. $30 \times 2 = 60$. The actual product of the multiplication will be about 60.

Second, arrange the factors to be multiplied vertically, with the ones digits in a single column, the tens digits in a second column, and so on.

$$\begin{array}{r} 34 \\ \times\ 2 \\ \hline \end{array}$$

Finally, to make the process clear, let's write it in expanded form.

$$\begin{array}{r} 34 \\ \times\ 2 \\ \hline \end{array} \qquad \begin{array}{r} 3 \text{ tens} + 4 \text{ ones} \\ \times\ 2 \\ \hline 6 \text{ tens} + 8 \text{ ones} = 60 + 8 = 68 \end{array}$$

Check: The guess 60 is roughly equal to the answer, 68.

Notice that when a single number multiplies a sum it forms a product with each addend in the sum. For example,

$$2 \times (3 + 4) = (2 \times 3) + (2 \times 4)$$

In the expanded multiplication above, the multiplier 2 forms a product with each addend in the sum (3 tens + 4 ones).

Write the following multiplication in expanded form.

$$\begin{array}{r} 28 \\ \times\ 3 \\ \hline \end{array}$$

Check your work in **30**.

30

Guess: $30 \times 3 = 90$

$$\begin{array}{r} 28 \\ \times\ 3 \\ \hline \end{array} \qquad \begin{array}{r} 2 \text{ tens} +\ \ 8 \text{ ones} \\ \times\ 3 \\ \hline 6 \text{ tens} + 24 \text{ ones} \end{array} = 6 \text{ tens} + 2 \text{ tens} + 4 \text{ ones}$$
$$= 8 \text{ tens} + 4 \text{ ones}$$
$$= 80 + 4$$
$$= 84$$

Check: 90 is roughly equal to 84.

Of course we do not normally use the expanded form. Instead we simplify the work like this:

■Step 1   ■Step 2   ■Step 3

```
  28                        28                        28
× 3   multiply ones       × 3                       × 3
  24   8 × 3 = 24           24   multiply             24   add partial
                           60   3 × 20 = 60           60     products
                                                      84   24 + 60 = 84
```

When you are certain about how to do this, you can take a short-cut and write:

```
  2
 28    3 × 8 = 24, write 4 and carry 20.
× 3    3 × 2 tens = 6 tens, 6 tens + 2 tens = 8 tens,
 84    write 8.
```

Try these problems to be certain you understand the process. Work all four problems using both the step-by-step and the short-cut methods.

(a)   43
     × 5

(b)   73
     × 4

(c)   29
     × 6

(d)   258
     ×   7

Check your work in **31.**

31

(a) Guess: $5 \times 40 = 200$

■Step 1   ■Step 2   ■Step 3

```
   43                         43                         43
 ×  5   5 × 3 = 15          ×  5   5 × 40 = 200        ×  5
 ────                       ────                       ────
   15                         15                         15
                             200                         200
                             ───                         ───
                                                         215
```

■Short-cut

```
   /
   43     5 × 3 = 15
 ×  5     5 × 4 tens = 20 tens
 ────
  215     20 tens + 1 ten = 21 tens
```

(b) Guess: $70 \times 4 = 280$

■Step 1   ■Step 2   ■Step 3

```
   73                         73                         73
 ×  4   3 × 4 = 12          ×  4   4 × 70 = 280        ×  4
 ────                       ────                       ────
   12                         12                         12
                             280                         280
                             ───                         ───
                                                         292
```

■Short-cut

```
   /
   73     4 × 3 = 12
 ×  4     4 × 7 tens = 28 tens
 ────
  292     28 tens + / ten = 29 tens
```

(c) Guess: $30 \times 6 = 180$

■Step 1   ■Step 2   ■Step 3   ■Short-cut

```
                                              5
   29            29            29            29        6 × 9 = 54
 ×  6          ×  6          ×  6          ×  6        6 × 2 tens = 12 tens
 ────          ────          ────          ────
   54            54            54            174       12 tens + 5 tens = 17 tens
                 120           120
                 ───           ───
                               174
```

(d) Guess: $300 \times 7 = 2100$

■Step 1   ■Step 2   ■Step 3

```
  258                        258                        258
 ×  7   7 × 8 = 56          ×  7   7 × 50 = 350        ×  7
 ────                       ────                       ────
   56                         56                         56
                             350                         350
                                                        1400   7 × 200 = 1400
```

(d) continued

■Step 4

```
   258
×    7
    56
   350
  1400
  1806   add
```

■Short-cut

```
     45
   258
×    7
  1806
```

Calculations involving two-digit multipliers are done in exactly the same way. Apply this method to this problem.

```
   57
×  24
```

The worked example is in **32**.

32

Guess: 60 × 20 = 1200

■Step 1

```
   57   Multiply by the
×  24   ones digit (4).
   28   4 × 7 = 28
  200   4 × 50 = 200
```

■Step 2

```
   57   Multiply by the
×  24   tens digit (2).
   28
  200
  140   20 × 7 = 140
 1000   20 × 50 = 1000
 1368   Add
```

Use the short-cut method illustrated on the next page to reduce the written work.

$$\begin{array}{r} \overset{\scriptstyle 1}{\underset{\scriptstyle 2}{}} \\ 57 \\ \times\ 24 \\ \hline 228 \\ 1140 \\ \hline 1368 \end{array}$$

$4 \times 7 = 28$, write 8, carry 2 tens.

$4 \times 50 = 200$, add carried 20 to get 220, write 22.

$20 \times 7 = 140$, write 40, carry 1 hundred.

$20 \times 50 = 1000$, add carried 100 to get 1100, write 11.

$1368 \leftarrow$ Add.

The zero in 1140 is usually omitted to save time.

Try these:

(a) $\begin{array}{r} 64 \\ \times\ 37 \end{array}$
    (b) $\begin{array}{r} 327 \\ \times\ 145 \end{array}$
    (c) $\begin{array}{r} 342 \\ \times\ 102 \end{array}$

Work each problem as shown above. Use the short-cut method if possible. Check your answers in **33**.

33

(a) Guess: $60 \times 40 = 2400$

$$\begin{array}{r} \overset{\scriptstyle 1}{\underset{\scriptstyle 2}{}} \\ 64 \\ \times\ 37 \\ \hline 448 \\ 1920 \\ \hline 2368 \end{array}$$

$7 \times 4 = 28$, write 8, carry 2 tens.

$7 \times 60 = 420$, add carried 20 to get 440, write 44.

$30 \times 4 = 120$, write 20, carry 1 hundred.

$30 \times 60 = 1800$, add carried 100 to get 1900, write 19.

$2368 \leftarrow$ Add.

(b) Guess: $300 \times 100 = 30,000$

$$\begin{array}{r} \overset{\scriptstyle 1\ 2}{\underset{\scriptstyle 1\ 3}{}} \\ 327 \\ \times\ 145 \\ \hline 1635 \\ 13080 \\ 32700 \\ \hline 47415 \end{array}$$

$5 \times 7 = 30$, write 5, carry 3 tens.

$5 \times 20 = 100$, add carried 30 to get 130, write 3, carry 1 hundred.

$5 \times 300 = 1500$, add carried 100 to get 1600, write 16.

$40 \times 7 = 280$, write 80, carry 2 hundreds.

$40 \times 20 = 800$, add carried 200 to get 1000, write 0, carry 10 hundreds.

$40 \times 300 = 12000$, add carried 1000 to get 13,000, write 13.

$100 \times 327 = 32700$

(c) Guess: $300 \times 100 = 30,000$

$$\begin{array}{r} 342 \\ \times\ 102 \\ \hline 684 \\ 000 \\ 34200 \\ \hline 34884 \end{array}$$

$2 \times 342 = 684$

$0 \times 342 = 000$

$100 \times 342 = 34200$

Be very careful when there are zeros in the multiplier; it is very easy to misplace one of those zeros. Do not skip any steps and be sure to guess 'n check.

Go to **34** for a set of practice problems on multiplication of whole numbers.

MULTIPLICATION SHORT CUTS

There are hundreds of quick ways to multiply various numbers. Most of them will confuse you more than help you unless you are already a math whiz. Here are a few that are easy to do and easy to remember.

1. To multiply by 10, annex a zero on the right end of the multiplicand. For example,

$$34 \times 10 = 340 \qquad\qquad 256 \times 10 = 2560$$

Multiplying by 100 or 1000 is similar.

$$34 \times 100 = 3400 \qquad\qquad 256 \times 1000 = 256000$$

2. To multiply by a number ending in zeros, carry the zeros forward to the answer. For example,

$\begin{array}{r} 26 \\ \times\ 20 \\ \hline \end{array}$ $\longrightarrow$ $\begin{array}{r} 26 \\ \times\ 20 \\ \hline 520 \end{array}$	Multiply 26×2 and attach the zero on the right. The product is 520.

$\begin{array}{r} 34 \\ \times 2100 \\ \hline \end{array}$ $\longrightarrow$ $\begin{array}{r} 34 \\ \times 2100 \\ \hline 34 \\ 68 \\ \hline 71400 \end{array}$

3. If both multiplier and multiplicand end in zeros, bring all zeros forward to the answer.

$\begin{array}{r} 230 \\ \times 200 \\ \hline \end{array}$ $\longrightarrow$ $\begin{array}{r} 230 \\ \times 200 \\ \hline 46000 \end{array}$ Attach three zeros to the product of 23×2.

$\begin{array}{r} 1000 \\ \times\ \ 100 \\ \hline 100,000 \end{array}$ This sort of multiplication is mostly a matter of counting zeros.

34

Problem Set 5: Multiplication

A. Multiply

7	9	7	7	6	9
6	8	8	9	8	6

8	6	6	9	8	8
9	7	9	7	7	6

B. Multiply

29	67	72	27	47	88	64	37	39	42
3	6	8	9	6	9	5	7	4	7

58	87	94	49	17	23	47	53	77	36
5	3	6	8	9	7	6	8	4	9

48	35	64	72	90	41	86	18	34	28
15	43	27	38	56	72	83	65	57	91

66	71	59	18	29	82	78	35	94	43
25	19	75	81	32	76	49	58	95	64

C. Multiply

305	145	3006	481	8043	765	809
123	516	125	203	37	502	47

1107	3706	4210	708	6401	684	319
98	102	304	58	773	45	708

2043	354	2008	923	563	8745
670	88	198	47	107	583

D. Brain Boosters

1. A room has 26 square yards of floor space. If carpeting costs $13 per square yard, what would it cost to carpet the room?

2. How many hours are there in a normal 365-day year?

3. If you manage to save $23 per week, how much money will you have in a year (52 weeks)?

4. A portable TV can be bought on credit for $25 down and twelve payments of $15 each. What is its total cost?

5. How many chairs are in a classroom with 27 rows of 14 seats each?

6. The Maharaja of Pourboy bought 16 Cadillacs, one for each of his wives, at a cost of $8165 each. What was the total cost of his purchase?

7. Multiply 123,456,789 by 8 and add 9 to the product. What is the total?

8. Complete each of these:

37	101	271	1221	4649	7373
× 3	× 11	× 41	× 91	× 239	× 1507

9. What is interesting about these three multiplications?

11313	11317	31311
× 10913	× 10917	× 30911

The answers to these problems are on page 224. When you have had the practice you need, turn to **35** to study the division of whole numbers or return to the preview on page 1 and use it to determine where to go next.

Division

35

Division is the reverse process for multiplication. It enables us to separate a given quantity into equal parts. The mathematical phrase $12 \div 3$ is read "twelve divided by three" and it asks us to separate a collection of 12 objects into 3 equal parts. The mathematical phrases

$$12 \div 3 \qquad 3\overline{)12} \qquad \frac{12}{3} \qquad 12/3$$

all represent division and are all read "twelve divided by three."
Perform this division:

$$12 \div 3 = \underline{\hspace{2cm}}$$

Turn to **36** to continue.

36

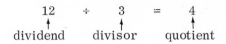

$$12 \quad \div \quad 3 \quad = \quad 4$$

dividend   divisor   quotient

In this division problem, the number being divided (12) is called the dividend, the number used to divide (3) is called the divisor, and the result of the division (4) is called the quotient. The word "quotient" comes from a Latin word meaning "how many times."

One way to perform division is to reverse the multiplication process.

$$24 \div 4 = \square \text{ means that } 4 \times \square = 24$$

If the one-digit multiplication tables are firmly set in your memory, you will recognize immediately that $\square = 6$ in this problem.

Try these.

$35 \div 7 = \underline{\hspace{1.5cm}}$ $42 \div 6 = \underline{\hspace{1.5cm}}$

$28 \div 4 = \underline{\hspace{1.5cm}}$ $56 \div 7 = \underline{\hspace{1.5cm}}$

$45 \div 5 = \underline{\hspace{1.5cm}}$ $18 \div 3 = \underline{\hspace{1.5cm}}$

$70 \div 10 = \underline{\hspace{1.5cm}}$ $63 \div 9 = \underline{\hspace{1.5cm}}$

$30 \div 5 = \underline{\hspace{1.5cm}}$ $72 \div 8 = \underline{\hspace{1.5cm}}$

Check your answers in **37**.

37

$35 \div 7 = 5$ $42 \div 6 = 7$

$28 \div 4 = 7$ $56 \div 7 = 8$

$45 \div 5 = 9$ $18 \div 3 = 6$

$70 \div 10 = 7$ $63 \div 9 = 7$

$30 \div 5 = 6$ $72 \div 8 = 9$

You should be able to do all of these quickly by working backward from the one-digit multiplication table.

How do we divide dividends that are larger than 9 × 9 and therefore not in the multiplication table? Obviously, we need a better procedure. One way to learn how many times the divisor divides the dividend is to subtract it repeatedly. For example, in 12 ÷ 3

12	9	6	3	3 is subtracted from 12 four times,
− 3	− 3	− 3	− 3	so that 12 ÷ 3 = 4.
9	6	3	0	

Try it. Perform the division 138 ÷ 23 using repeated subtraction. Check your answer in **38.**

AVERAGES

The <u>average</u> of a set of numbers is the single number that best represents the whole set. One simple kind of average is the arithmetic average or arithmetic mean defined as

$$\text{arithmetic average} = \frac{\text{sum of measurements}}{\text{number of measurements}}$$

For example, the average weight of the five middle linemen on our college football team is figured below.

$$\text{average weight} = \frac{215 \text{ lb} + 235 \text{ lb} + 224 \text{ lb} + 212 \text{ lb} + 239 \text{ lb}}{5}$$

$$= \frac{1125 \text{ lb}}{5} = 225 \text{ lb}$$

Try these problems for practice.

1. In a given week a student worked in the college library for the following hours each day: Monday, 3 hours; Tuesday, 3 hours; Wednesday, $2\frac{1}{2}$ hours; Thursday, $3\frac{1}{4}$ hours; Friday, $2\frac{1}{4}$ hours; and Saturday, 4 hours. What average amount of time does the student work per day?
2. On four weekly quizzes in her history class, a student scores 84, 74, 92, and 88 points. What is her average score?
3. A salesman sells the following amounts in successive weeks: $647.20, $705.17, $1205.65, $349.34, and $409.89. What is his average weekly sales?

The answers to these problems are on page 224.

38

$$138 \div 23$$

138	115	92	69	46	23
- 23	- 23	- 23	- 23	- 23	- 23
115	92	69	46	23	0

23 may be subtracted from 138 six times, therefore $138 \div 23 = 6$.

We could also divide by simply guessing. For example, to find $245 \div 7$ by guessing we might go through a mental conversation with ourselves something like this:

> "7 into 245 goes how many times?"
> "Lots."
> "How many? Pick a number."
> "Maybe 10."
> "Let's try it: $7 \times 10 = 70$, so 10 is much too small. Try a larger number."
> "How about 50? Does 7 go into 245 about 50 times?"
> "Well, $7 \times 50 = 350$, which is larger than 245. Try again."
> "I'm getting tired. Will 30 do it?"
> "Well, $7 \times 30 = 210$. That is quite close to 245. Try something a little larger than 30."
> "31? 32? . . ."

Sooner or later the tired little guesser in your head will arrive at 35 and find that $7 \times 35 = 245$, so

$$245 \div 7 = 35$$

With pure guessing even a simple problem could take all afternoon. We need a short-cut. The best division process combines one-digit multiplication, repeated subtraction, and educated guessing. For example, in the problem $96 \div 8$, start with a guess: the answer is about 10, since $8 \times 10 = 80$.

■Step 1

Arrange the divisor and dividend horizontally.

tens column ↘ ↙ ones column

$8\overline{)96}$

Can 8 be subtracted from 9? Yes, once. Write 1 in the tens column and place a zero in the ones column. 10 is your first guess at the quotient.

$$\begin{array}{r} 10 \\ 8\overline{)96} \end{array}$$

■Step 2

Multiply $8 \times 10 = 80$.
Subtract $96 - 80 = 16$.

$$\begin{array}{r} 10 \\ 8\overline{)96} \\ -80 \\ \hline 16 \end{array}$$

■Step 3

Use 16 as the new dividend. Can 8 be
subtracted from 1? No. Write a zero
in the tens column above. Can 8 be
subtracted from 16? Yes, twice. Write
a 2 in the ones column above.

$$\begin{array}{r} 02 \\ 10 \\ 8\overline{)96} \\ -80 \\ \hline 16 \end{array}$$

■Step 4

$$\begin{array}{r} 02 \\ 10 \\ 8\overline{)96} \\ -80 \\ \hline 16 \end{array}$$

Multiply $8 \times 2 = 16$.
Subtract $16 - 16 = 0$.

$$\begin{array}{r} 16 \\ -16 \\ \hline 0 \end{array}$$

The quotient is the sum of the numbers in the answer space ($10 + 2 = 12$)
so that $96 \div 8 = 12$. Always check your answer: $8 \times 12 = 96$.
Now you try one:

$112 \div 7 = $ _____

Work this problem using the method shown above. Then go to **39**.

39

$112 \div 7$ 	 Guess: $7 \times 10 = 70$ and $7 \times 20 = 140$ so the
answer is between 10 and 20.

■Step 1

Can 7 be subtracted from 1? No. Write a
zero above the 1 in the hundreds column.
Can 7 be subtracted from 11? Yes, once.
Write a 1 in the tens column. Place a
zero in the ones column.

$$\begin{array}{r} 010 \\ 7\overline{)112} \end{array}$$

■Step 2

Multiply 7 × 10 = 70 and subtract
112 – 70 = 42.

■Step 3

Use 42 as the new dividend and repeat the
process: 7 × 6 = 42, 42 – 42 = 0.
Quotient = 10 + 6 = 16, remainder = 0.

Check: 7 × 16 = 112

$$\begin{array}{r} 6 \\ 010 \\ 7\overline{)112} \\ -70 \\ \hline 42 \\ -42 \\ \hline \end{array}$$

When you get comfortable with this process, you can omit writing zeros
and your work will look like this:

$$\begin{array}{r} 16 \\ 7\overline{)112} \\ 7 \\ \hline 42 \\ 42 \\ \hline 0 \end{array}$$

7 into 11 once, write 1.
11 – 7 = 4, bring down the 2.
7 into 42, 6 times, write 6.
7 × 6 = 42, 42 – 42 = 0.

Here are a few problems for practice in division.

(a) 976 ÷ 8   (b) 3174 ÷ 6   (c) 204 ÷ 6

Check your work in **40.**

SPECIAL DIVISORS

A few divisors require special attention. Remember that, for
any two numbers a and b, a ÷ b = □ means that b × □ = a. That
is, 21 ÷ 3 = 7 means 3 × 7 = 21.

1. If any number is divided by one, the quotient is the original
    number.
    $$6 \div 1 = 6 \text{ and } \frac{6}{1} = 6 \text{ because } 6 \times 1 = 6$$

2. If any number is divided by itself, the quotient is one.
    $$6 \div 6 = 1 \text{ and } \frac{6}{6} = 1 \text{ because } 6 \times 1 = 6$$

3. If zero is divided by any nonzero number, the quotient is zero.
    $$0 \div 6 = 0 \text{ and } \frac{0}{6} = 0 \text{ because } 6 \times 0 = 0$$

4. If any number is divided by zero, the quotient is not defined in mathematics. If $6 \div 0 = \square$ then $0 \times \square = 6$, but 0 times any number equals zero. There is no number $\square$ that will make this equation true.

5. The fraction $\dfrac{0}{0}$ is never used because it can have any value whatever. If $\dfrac{0}{0} = \square$ then $0 = 0 \times \square$ and this equation is true for any value of $\square$.

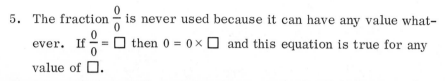

40

(a) Guess: $8 \times 100 = 800$; the answer is something close to 100.

Step 1: 8 goes into 9 once, write 1.

Step 2: $8 \times 1 = 8$, subtract $9 - 8 = 1$.

Step 3: Bring down 7. 8 goes into 17 twice, write 2.

Step 4: $8 \times 2 = 16$, subtract $17 - 16 = 1$.

Step 5: Bring down 6. 8 goes into 16 twice, write 2.

Step 6: $8 \times 2 = 16$, subtract $16 - 16 = 0$.

Check: $8 \times 122 = 976$

$$
\begin{array}{r}
122 \\
8\overline{)976} \\
-8 \\
\hline
17 \\
-16 \\
\hline
16 \\
-16 \\
\hline
0
\end{array}
$$

Quotient = 122

(b) Guess: $6 \times 500 = 3000$; the answer is roughly 500.

Step 1: 6 into 3? No. 6 goes into 31 five times, write 5.

Step 2: $6 \times 5 = 30$, subtract $31 - 30 = 1$.

Step 3: Bring down 7. 6 goes into 17 twice, write 2.

Step 4: $6 \times 2 = 12$, subtract $17 - 12 = 5$.

Step 5: Bring down 4. 6 goes into 54 nine times, write 9.

Step 6: $6 \times 9 = 54$, subtract $54 - 54 = 0$.

Check: $6 \times 529 = 3174$

$$
\begin{array}{r}
529 \\
6\overline{)3174} \\
-30 \\
\hline
17 \\
-12 \\
\hline
54 \\
-54 \\
\hline
0
\end{array}
$$

Quotient = 529

(c) Guess: $6 \times 30 = 180$, so the quotient is about 30.

$$
\begin{array}{r}
34 \\
6\overline{)204} \\
-18 \\
\hline
24 \\
24 \\
\hline
0
\end{array}
$$

Step 1: 6 into 2? No. 6 goes into 20 three times, write 3.

Step 2: $6 \times 3 = 18$, subtract $20 - 18 = 2$.

Step 3: Bring down 4. 6 goes into 24 four times, write 4.

Quotient = 34

Step 4: $6 \times 4 = 24$, subtract $24 - 24 = 0$.

Check: $6 \times 34 = 204$

Now try a problem using a two-digit divisor.

$5084 \div 31 =$ _____

The procedure is the same as above. Check your answer in **41**.

41

$5084 \div 31$

Guess: This is roughly the same as $500 \div 3$ or about 200. The quotient will be about 200.

Step 1: 31 into 5? No. 31 goes into 50 once, write 1.

$$
\begin{array}{r}
164 \\
31\overline{)5084} \\
-31 \\
\hline
198 \\
-186 \\
\hline
124 \\
124 \\
\hline
0
\end{array}
$$

Step 2: $31 \times 1 = 31$, subtract $50 - 31 = 19$.

Step 3: Bring down 8. 31 into 198? (That is about the same as 3 into 19.) Yes, 6 times, write 6.

Step 4: $31 \times 6 = 186$, subtract $198 - 186 = 12$.

Quotient = 164

Step 5: Bring down 4. 31 into 124? (That is about the same as 3 into 12.) Yes, 4 times, write 4.

Step 6: $31 \times 4 = 124$, subtract $124 - 124 = 0$.

Check: $31 \times 164 = 5084$

Notice that in step 3 it is not at all obvious how many times 31 will go into 198. Again, you must make an educated guess and check your guess as you go along. If you guess 'n check on every problem you will always get the correct answer.

So far, we have looked only at division problems that "come out even." In these problems the remainder is zero. Obviously not all division problems are of this kind. What would you do with these?

(a) 59 ÷ 8 = _____   (b) 341 ÷ 43 = _____

(c) 7528 ÷ 37 = _____

Look in **42** for our answers.

42

(a)   7   The quotient is 7 with a remainder of 3.
       8 ⟌59

       −56
        3

(b)   7   Your first guess would probably be that 43 goes into 341
      43 ⟌341   eight times (try 4 into 34) but 43 × 8 = 344 which is larger
       −301   than 341. The quotient is therefore 7 with a remainder of
        40   40.

(c)   203
      37 ⟌7528

       −74
        128 ◄—At this point notice that 37 cannot be subtracted from 12.
       −111   Write a zero in the answer space and bring down the 8.
         17   The quotient is 203 and the remainder is 17.

Now turn to **43** for a set of practice problems on division of whole numbers.

43

Problem Set 6: Division

A. Divide

       63 ÷ 7 = _____   84 ÷ 7 = _____   92 ÷ 8 = _____

$56 \div 8 =$ _____ $72 \div 0 =$ _____ $65 \div 5 =$ _____

$37 \div 5 =$ _____ $45 \div 9 =$ _____ $71 \div 7 =$ _____

$7 \div 1 =$ _____ $6 \div 6 =$ _____ $13 \div 0 =$ _____

$\dfrac{32}{4} =$ _____ $\dfrac{18}{3} =$ _____ $\dfrac{28}{7} =$ _____

$\dfrac{42}{6} =$ _____ $\dfrac{54}{9} =$ _____ $\dfrac{63}{7} =$ _____

B. Divide

$245 \div 7 =$ _____ $369 \div 9 =$ _____ $167 \div 7 =$ _____

$126 \div 3 =$ _____ $228 \div 4 =$ _____ $232 \div 5 =$ _____

$310 \div 6 =$ _____ $360 \div 8 =$ _____ $337 \div 3 =$ _____

$\dfrac{132}{3} =$ _____ $\dfrac{147}{7} =$ _____ $\dfrac{216}{8} =$ _____

$7\overline{)364}$ $6\overline{)222}$ $8\overline{)704}$

$5\overline{)625}$ $4\overline{)201}$ $9\overline{)603}$

C. Divide

$322 \div 14 =$ _____ $357 \div 17 =$ _____ $382 \div 19 =$ _____

$407 \div 13 =$ _____ $936 \div 24 =$ _____ $502 \div 10 =$ _____

$700 \div 28 =$ _____ $701 \div 36 =$ _____ $730 \div 81 =$ _____

$\dfrac{451}{11} =$ _____ $\dfrac{901}{17} =$ _____ $\dfrac{989}{23} =$ _____

$31\overline{)682}$ $43\overline{)507}$ $61\overline{)732}$

$12\overline{)408}$ $33\overline{)303}$ $13\overline{)928}$

D. Divide

$2001 \div 21 =$ _____ $3328 \div 32 =$ _____ $2016 \div 21 =$ _____

$3536 \div 17 =$ _____ $1000 \div 7 =$ _____ $5029 \div 47 =$ _____

$2000 \div 9 =$ _____ $1881 \div 11 =$ _____ $2400 \div 75 =$ _____

$7 \overline{)7000}$ $\qquad$ $14 \overline{)4275}$ $\qquad$ $27 \overline{)8405}$

$71 \overline{)6005}$ $\qquad$ $31 \overline{)3105}$ $\qquad$ $53 \overline{)6307}$

$231 \overline{)14091}$ $\qquad$ $411 \overline{)42020}$ $\qquad$ $603 \overline{)48843}$

E. Brain Boosters

1. The longest footrace ever held started in New York and ended in Los Angeles in 1929. If the winner covered about 3680 miles in 525 hours, what was his average speed?

2. Sally earns $6084 per year after taxes. What monthly paycheck should she expect?

3. The weights of the seven linemen on the Gnu U football team are: 210, 215, 245, 217, 220, 227, and 115 lb (he is very fast). What is the average weight per man for the line? (Add up and divide by 7.)

4. Suppose your little Honda sports car travels 420 miles on 11 gallons of gasoline. How many miles to the gallon are you getting?

5. Last week the People's Hickey Company sold 213 hickeys for a total of $10,863. What does one hickey cost?

6. Typing at the rate of 72 words per minute, how long would it take to type 1800 words?

7. The planet Pluto travels once around the sun in approximately 90,464 days. If one earth year is equal to 365 days, how many earth years is a Pluto year?

8. The noise of an explosion travels 6200 meters through the air in 18 seconds. What is the speed of this sound? (Your answer should be in meters per second.)

The answers to these problems are on page 224. When you have had the practice you need, continue by going to the factoring of whole numbers in **44** or by returning to the preview for this chapter on page 1.

Factors and Factoring

44

The symbols 6, Vi, and ⊞⊦ι are all names for the number six. We can also write any number in terms of arithmetic operations involving other numbers. For example, (4 + 2), (7 − 1), (2 × 3), and (18 ÷ 3) are also ways of writing the number six. It is particularly useful to be able to write any whole number as a product of other numbers. If we write

$$6 = 2 \times 3$$

2 and 3 are called the factors of 6. Of course we could write

$$6 = 1 \times 6$$

and see that 1 and 6 are also factors of 6, but this does not tell us anything new about the number 6. The factors of 6 are 1, 2, 3, and 6.

What are the factors of 12? (Choose an answer.)

(a) 2, 3, 4, and 6   Go to **45.**
(b) 1, 2, 3, 4, 6, and 12   Go to **46.**
(c) 0, 1, 2, 3, 4, 6, and 12   Go to **47.**

45

Not quite. Any two whole numbers whose product is 12 are factors of 12. It is easy to see that

$$1 \times 12 = 12$$

Therefore, 1 and 12 are factors of 12.

Return to **44** and choose a better answer.

46

Right you are.

$$1 \times 12 = 12 \qquad\qquad 2 \times 6 = 12 \qquad\qquad 3 \times 4 = 12$$

are all ways of writing 12 as the product of two numbers. Therefore, 1, 2, 3, 4, 6, and 12 are all factors of 12.

Any number is evenly divisible by its factors; that is, every factor divides the number with zero remainder. For example,

$$12 \div 1 = 12 \qquad\qquad 12 \div 4 = 3$$
$$12 \div 2 = 6 \qquad\qquad 12 \div 6 = 2$$
$$12 \div 3 = 4 \qquad\qquad 12 \div 12 = 1$$

List the factors of these numbers:

(a) 18   (b) 20

(c) 24   (d) 48

Check your answers in **48.**

47

Not correct. Zero is never a factor of any number. There is no number ☐ such that

$$0 \times \square = 12$$

The product of 0 and any number is always 0.
Return to **44** and choose a better answer.

48

    (a) The factors of 18 are 1, 2, 3, 6, 9, and 18.
    (b) The factors of 20 are 1, 2, 4, 5, 10, and 20.
    (c) The factors of 24 are 1, 2, 3, 4, 6, 8, 12, and 24.
    (d) The factors of 48 are 1, 2, 3, 4, 6, 8, 12, 16, 24, and 48.

For some numbers the only factors are 1 and the number itself. For example, the factors of 7 are 1 and 7 because

$$1 \times 7 = 7$$

There are no other numbers that divide 7 with remainder zero. Such numbers are known as <u>prime</u> <u>numbers</u>. A prime number is one for which there are no factors other than 1 and the prime number itself.
    Here is a list of the first few prime numbers:

2	3	5	7	11
13	17	19	23	29
31	37	41	43	47

Notice that 1 is not listed. All prime numbers have two distinct, unequal factors: 1 and the number itself. The number 1 has only one factor—itself. The number 1 is not a prime.
    There are twenty-five prime numbers less than 100, 168 less than 1000, and no limit to the total number. Mathematicians have tried for centuries to find a simple pattern that would enable them to write down the primes in order and predict if any given number is a prime. As yet no one has succeeded.
    How then does one determine if some given number is a prime? There is no magic way to decide. You must divide the number by each whole number in order, starting with 2. If the division has no remainder, the original number is not a prime. Continue dividing until the quotient obtained is less than the divisor.
    For example, is 53 a prime number? Try to work it out, then turn to **49.**

49

To decide if 53 is a prime we must perform a series of divisions:

$$\frac{26}{2\,\overline{\smash{\big)}\,53}} \quad \text{remainder} = 1 \qquad\qquad \frac{10}{5\,\overline{\smash{\big)}\,53}} \quad \text{remainder} = 3$$

$$\frac{17}{3\,\overline{\smash{\big)}\,53}} \quad \text{remainder} = 2 \qquad\qquad \frac{8}{6\,\overline{\smash{\big)}\,53}} \quad \text{remainder} = 5$$

$$\frac{13}{4\,\overline{\smash{\big)}\,53}} \quad \text{remainder} = 1 \qquad\qquad \frac{7}{7\,\overline{\smash{\big)}\,53}} \quad \text{remainder} = 4$$

We can stop the search for a divisor here because dividing by 8 gives a quotient (6) less than the divisor (8). All divisors produce a nonzero remainder, therefore 53 is a prime.

Notice that it is really not necessary to test divide by 4 or 6. If the number is not evenly divisible by 2, it cannot be evenly divided by 4 or 6 because these are multiples of 2. In testing for primeness we need test divide only by primes. This will save time, but you must have the first few primes memorized. For most of your work it will be sufficient if you remember the first eight or ten primes listed in **48**. A study card listing primes to be memorized is included in the back of this book (page 243). Use it.

Which of the following numbers are prime?

(a) 103   (b) 114   (c) 143   (d) 223

(e) 289   (f) 449   (g) 527   (h) 667

Test divide by primes as shown above, then check your answers in **50**.

THE SIEVE OF ERATOSTHENES

More than 2000 years ago, Eratosthenes, a Greek geographer-astronomer, devised a way of locating primes that is still the most effective known. His procedure separates the primes out of the set of all whole numbers. Here is one version of what he did. First, arrange the whole numbers in six columns starting with 2 as shown. Second, circle the prime 2 and cross out all the multiples of 2; circle the next number (3) and cross out all multiples of 3; circle the next remaining number (5) and cross out all multiples of 5; and so on. The circled numbers remaining are the primes.

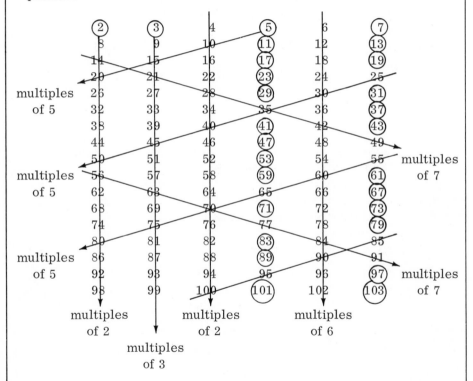

Mathematicians call this procedure a "sieve" because it is a way of sorting out or separating the primes from the other whole numbers.

Notice that all primes greater than 3 are either in the 5 or 7 column. They are either one less or one more than a multiple of 6. Pairs of primes separated by one integer (such as 5 and 7 or 11 and 13) are called <u>twin primes</u>. Can you find any other interesting patterns?

50

(a) 103 ÷ 2 = 51   remainder 1   103 ÷ 3 = 34   remainder 1
   103 ÷ 5 = 20   remainder 3   103 ÷ 7 = 14   remainder 5
   103 ÷ 11 = 9   remainder 4
   At this point there is no need to continue because the quotient (9) is less than the divisor (11).   103 is a prime.

(b) 114 ÷ 2 = 57   no remainder
   Because 114 is evenly divisible by 2, it is not a prime.

(c) 143 ÷ 11 = 13   no remainder
   Because 143 is evenly divisible by 11 and 13, it is not a prime.

(d) 223 is a prime. Test divide it by 2, 3, 5, 7, and 11.

(e) 289 = 17 × 17, so 289 is not a prime.

(f) 449 is a prime. Test divide it by 2, 3, 5, 7, 11, 13, 17, and 19.

(g) 527 = 17 × 31, so 527 is not a prime.

(h) 667 = 23 × 29, so 667 is not a prime.

● The prime factors of any numbers are those factors that are prime numbers.

The prime factors of 6 are 2 and 3. The prime factors of 21 are 3 and 7. The factors of 42 are 6 and 7, but the prime factors of 42 are 2, 3, and 7. The number 6 is not a prime factor; it is not a prime. To factor a number means to find its prime factors. Finding the prime factors of a number is a necessary skill if you want to learn how to add and subtract fractions.

What are the prime factors of 30?

(a) 5 × 6 = 30   Go to **51**.
(b) 2, 3, 5, 6, 10, 15   Go to **52**.
(c) 2, 3, 5   Go to **53**.

51

You goofed on this one. The numbers 5 and 6 are factors of 30. Their product equals 30. But they are not both prime factors. Prime factors must be prime numbers.

Return to **50** and choose a different answer.

52

These are some of the factors of 30, but not all of them are prime. Return to **50** and choose the set of prime factors.

53

Right! The prime factors of 30 are 2, 3, and 5.

Prime numbers are especially interesting because any number can be written as the product of primes in only one way. $6 = 2 \times 3$ and $6 = 3 \times 2$ count as only one way of writing 6 as a product of primes. The order is not important. The key idea is that 2 and 3 are the <u>only</u> primes whose product is 6.

$9 = 3 \times 3$	This is a product of primes.
$10 = 2 \times 5$	So is this.
$11 = 11$	An easy one, 11 is itself a prime. We would not write 11×1 because 1 is not a prime . . . remember?
$12 = 2 \times 2 \times 3$	There are several ways to write 12 in terms of its factors ($12 = 2 \times 6$ or $12 = 3 \times 4$) but only one way with primes. Notice that the prime factor 2 must appear <u>twice</u>.

Write 60 as the product of its prime factors.

60 = _____

Check your work in **54**.

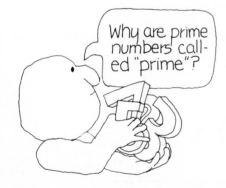

Why are prime numbers called "prime"?

The Latin word <u>primus</u> means first in importance and the primes are the important main ingredients of numbers. Every number is either a prime or a product of primes.

54

$60 = 2 \times 2 \times 3 \times 5$

Write them in any order you like, $2 \times 5 \times 3 \times 2$ or $5 \times 2 \times 3 \times 2$ or whatever. This is the <u>only</u> set of primes whose product is 60. It is this property of prime numbers that prompted the Greek mathematicians thirty centuries ago to call them "primes"—the "first" numbers from which the rest could be built.

Being able to write any number as a product of primes is a valuable skill. You will need this skill when you work with fractions in Chapter 2.

How can we rewrite a number in terms of its prime factors? Let's work through an example. Find the prime factors of 315. First, divide by the primes in order starting with 2.

2|315   2 is not a factor because it does not divide the number evenly. Bring the number down and try the next prime.

3|315   3 is a factor. Write it to the side ⟶ $\boxed{3}$
3|105   and divide by 3 again.
3|35

3 is a factor again. Write it to the side ⟶ $\boxed{3}$
and divide by 3 again.

3 does not divide into 35 evenly. Bring down 35 and try the next prime.

5|35   5 is a factor. Write it to the side. ⟶ $\boxed{5}$
7

7 is a factor. Write it to the side. ⟶ $\boxed{7}$

So 315 = 3 × 3 × 5 × 7 written as a product of prime factors. Finally, check this multiplication to be certain that you have missed no factors.

$$3 \times 3 \times 5 \times 7 = 9 \times 5 \times 7 = 45 \times 7 = 315$$

Find the prime factors of 693. Check your work in **55.**

THE PRIMES LESS THAN 100

2	3	5	7	11
13	17	19	23	29
31	37	41	43	47
53	59	61	67	71
73	79	83	89	97

THE FACTOR TREE

A very helpful way to think about factors is in terms of a factor tree. For example, factor 1764.

First, divide by the smallest prime, 2.
1764 ÷ 2 = 882
Write down the 2 and the quotient 882.

Second, divide by 2 again.
882 ÷ 2 = 441
On a new row write down both 2s and the quotient 441.

Third, when 2 will no longer divide the last quotient evenly, divide by the next largest prime, 3 . . . and so on. Stop when you find a prime quotient.

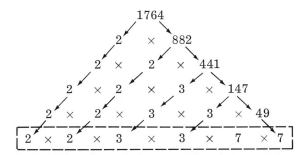

At each level of the tree the product of the horizontal numbers is equal to the original number to be factored. The last row gives the prime factors.

55

Your work should look something like this:

2 | 693   2 does not divide 693 evenly.

3 | 693   693 ÷ 3 = 231 —————————————————→ ③
3 | 231   231 ÷ 3 = 77 —————————————————→ ③
  3 | 77

3 does not divide 77 evenly.

5 | 77   5 does not divide 77 evenly.

7 | 77   77 ÷ 7 = 11 —————————————————→ ⑦
  11

11 is a prime. —————————————————→ ⑪

693 = 3 × 3 × 7 × 11   Check: 3 × 3 × 7 × 11 = 9 × 7 × 11
= 63 × 11 = 693

Use this method to find the prime factors of these numbers:

(a) 570   (b) 792   (c) 945

Our work is in **56**.

56

(a) 2 | 570   570 ÷ 2 = 285 ———→ ②
    2 | 285

not a divisor

3 | 285   285 ÷ 3 = 195 ———→ ③
  3 | 95

not a divisor

5 | 95   95 ÷ 5 = 19 ———→ ⑤
  19   19 is a prime ———→ ⑲

570 = 2 × 3 × 5 × 19

(b) 2 | 792   792 ÷ 2 = 396 ———→ ②
    2 | 396   396 ÷ 2 = 198 ———→ ②
    2 | 198   198 ÷ 2 = 99 ———→ ②
    2 | 99

not a divisor

3 | 99   99 ÷ 3 = 33 ———→ ③
  3 | 33   33 ÷ 3 = 11 ———→ ③
    11   11 is a prime ———→ ⑪

792 = 2 × 2 × 2 × 3 × 3 × 11

(c) 2 |945   not a divisor

3 |945   945 ÷ 3 = 315 ——————▶ ③
3 |315
3 |105   315 ÷ 3 = 105 ——————▶ ③
3 |35   105 ÷ 3 = 35 ——————▶ ③

not a divisor

5 |35   35 ÷ 5 = 7 ——————▶ ⑤
7   7 is a prime ——————▶ ⑦

945 = 3 × 3 × 3 × 5 × 7

It is possible to tell at a glance, without actually dividing, if any number is evenly divisible by 2, 3, or 5. Knowing how to do so can save you a bit of work. Use these divisibility rules.

- Any number is evenly divisible by 2 if its ones digit is 2, 4, 6, 8, or 0.
  Examples: 12 is divisible by 2, it ends in 2.
             46 is divisible by 2, it ends in 6.
             7498 is divisible by 2, it ends in 8.

- Any number is evenly divisible by 3 if the sum of its digits is divisible by 3.
  Examples: 18 is divisible by 3 since 1 + 8 = 9 and 9 is divisible by 3.
             471 is divisible by 3 since 4 + 7 + 1 = 12 and 12 is divisible by 3.
             72,954 is divisible by 3 since 7 + 2 + 9 + 5 + 4 = 27 and 27 is divisible by 3.
             215 is not divisible by 3 since 2 + 1 + 5 = 8, and 8 is not divisible by 3.

- Any number is evenly divisible by 5 is its ones digit is 5 or 0.
  Examples: 25 is divisible by 5, it ends in 5.
             370 is divisible by 5, it ends in 0.
             73,495 is divisible by 5, it ends in 5.

There are divisibility rules for other numbers (shown in the box on the next page, if you're interested), but the few above are all you really need to remember. Use these rules to decide which of the following numbers are divisible by 2, 3, or 5. Do all work mentally. Check ✓ those evenly divisible by 2. Mark with an ✗ those evenly divisible by 3. Circle those evenly divisible by 5.

16	23	27	39	45	111	132
330	335	372	453	498	785	921
1017	2111	73,908		123,456	4,271,305	

Turn to **57** to check your answers.

DIVISIBILITY RULES

Consider the following telephone number: 2616130. In a few seconds and without using pencil and paper, can you show that it is evenly divisible by 2, 3, 5, 6, 10, and 11 but not by 4, 7, 8, or 9? The trick is to use the following divisibility rules.

● 2   Any number is divisible by 2 if its ones digit is 2, 4, 6, 8, or 0.
Example: 14, 96, and 378 are all divisible by 2.

● 3   Any number is evenly divisible by 3 if the sum of its digits is divisible by 3.
Example: 672 is divisible by 3 since $6 + 7 + 2 = 15$ and 15 is divisible by 3.

● 4   Any number is evenly divisible by 4 if the number formed by its two rightmost digits is divisible by 4.
Example: 716 is divisible by 4 since 16 is divisible by 4.

● 5   Any number is evenly divisible by 5 if its ones digit is 0 or 5.
Example: 35, 90, and 1365 are all divisible by 5.

● 6   Any number is divisible by 6 if it is divisible by both 2 and 3.
Example: 822 is divisible by 6 since its ones digit is 2 and $8 + 2 + 2 = 12$ which is divisible by 3.

● 8   Any number is divisible by 8 if its last three digits are divisible by 8.
Example: 1160 is divisible by 8 since 160 is divisible by 8.

● 9   Any number is divisible by 9 if the sum of its digits is divisible by 9.
Example: 9243 is divisible by 9 since $9 + 2 + 4 + 3 = 18$ which is divisible by 9.

● 10   Any number whose ones digit is 0 is divisible by 10.
Example: 60, 210, and 19,830 are all divisible by 10.

There are no really simple rules for 7, 11, or 13. Here are the least complicated rules known.

● 7   Divide the number in question by 50. Add the quotient and remainder. The original number is divisible by 7 if the sum of the quotient and the remainder is divisible by 7.
Example: 476 ÷ 50 = 9, remainder = 26.
Add 9 + 26 = 35.
35 is divisible by 7, therefore 476 is also divisible by 7.

●11   Divide the number in question by 100. Add the quotient and remainder. The original number is divisible by 11 if the sum of the quotient and the remainder is divisible by 11.
Example: 1562 ÷ 100 = 15, remainder = 62.
Add 15 + 62 = 77.
77 is divisible by 11, therefore 1562 is also divisible by 11.

●13   Proceed as with 11, but divide by 40.
Example: 1170 ÷ 40 = 29, remainder 10.
Add 29 + 10 = 39.
39 is divisible by 13, therefore 1170 is also divisible by 13.

57

✓ means divisible by 2.   ✗ means divisible by 3.   Circle means divisible by 5.

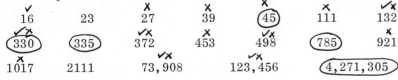

Now you should be ready for some practice problems on factoring. Turn to **58** and continue.

58

Problem Set 7: Factoring

A. List the prime factors of the following numbers:

12 _____   16 _____   14 _____

18 _____   24 _____   20 _____

26 _____   31 _____   32 _____

36 _____   39 _____   42 _____

56 _____   81 _____   121 _____

B. Write the following as products of primes:

96 = _____   84 = _____   136 = _____

170 = _____   252 = _____   256 = _____

288 = _____   390 = _____   468 = _____

546 = _____   980 = _____   1369 = _____

1363 = _____   1548 = _____   3149 = _____

C. Circle the primes among the following numbers:

6   2   5   9   1   14   3   31   21   23

37   39   53   15   26   19   67   61   63   72

27   91   89   87   17

D. Which of the following numbers are divisible by 2? Which by 3?
Which by 5? (Do it in your head.)

9   12   4   231   45   144   17   261   1044   1390

72   81   102   2808   2088   8280   8802   11   111   1111

E. Brain Boosters

1. A perfect number is one that is the sum of its divisors, not
counting itself. For example, 6 = 1 + 2 + 3 and 1, 2, and 3 are
divisors of 6. Show that 28, 496, and 8128 are also perfect numbers.

2. The numbers 220 and 284 are called "amicable" or "friendly"
numbers. It has been believed for hundreds of years that you
can maintain a friendship by exchanging gifts, each related to
one of these numbers. For example, in the Bible, Genesis
32:14, Jacob gave his brother Esau 220 goats and 220 sheep in
an attempt to appease him and gain his friendship.

a. What property of these numbers makes them special?
(Hint: add up the divisors of 284, not counting itself.
Repeat for 220.)

b. About 400 such pairs are known, each made up of the parts
of the other. Show that 2620 and 2924 are also a friendly
pair.

3. Complete this magic square so that all rows, columns, and diagonals add to the same sum. What is special about the numbers in this square?

67	1	43
13		61
31	73	7

4. Consecutive numbers are whole numbers that differ by one. For example, 2 and 3 are consecutive, so are 115 and 116, or 734, 735, and 736. The product of three consecutive numbers is always divisible by what three numbers? Can you see why this is so?

5. Which of the following are primes? What are the prime factors of the others?

   1   11   111   1,111   11,111   111,111

6. Fill in the missing digits marked with an asterisk (*):

```
   63*          *752           3*4
   *75          3*58           92*
 + 253          49*5         + *05
 ----          + 240*         ----
   *3*6        ------          1945
                *5788
```

The answers to these problems are on page 225. When you have had the practice you need continue by turning to **59** to study exponents and square roots or return to the preview on page 1.

Exponents and Square Roots

59

When the same number appears many times as a factor, writing the product may become monotonous, tiring, and even inaccurate. It is easy, for example, to miscount the twos in

$$131,072 = 2 \times 2 \times 2 \times 2 \times 2 \times 2 \times 2 \times 2 \times 2 \times 2 \times 2 \times 2 \times 2 \times 2 \times 2 \times 2 \times 2$$

or the tens in

$$100,000,000,000 = 10 \times 10 \times 10 \times 10 \times 10 \times 10 \times 10 \times 10 \times 10 \times 10 \times 10$$

Products of this sort are usually written in a shorthand form as 2^{17} or 10^{11}. In this <u>exponential form</u> the raised number 17 indicates the number of times 2 is to be used as a factor. For example,

$$2 \times 2 = 2^2 \qquad \text{product of } \underline{\text{two}} \text{ factors of 2}$$
$$2 \times 2 \times 2 = 2^3 \qquad \text{product of } \underline{\text{three}} \text{ factors of 2}$$
$$\underbrace{2 \times 2 \times 2 \times 2}_{\text{four 2s}} = 2^4 \qquad \text{product of } \underline{\text{four}} \text{ factors of 2}$$

Write $3 \times 3 \times 3 \times 3 \times 3$ in exponential form.

$$3 \times 3 \times 3 \times 3 \times 3 = \underline{\hspace{2cm}}$$

Check your answer in **60.**

60

$$\underbrace{3 \times 3 \times 3 \times 3 \times 3}_{\text{five factors of 3}} = 3^5$$

In this expression 3 is called the <u>base</u> and 5 is called the <u>exponent</u>. The exponent 5 tells you how many times the base 3 must be used as a factor in the product.

Multiply the factors in 4^3.

$$4^3 = \underline{\hspace{2cm}}$$

Pick an answer:

    (a) 12   Go to **61.**
    (b) 64   Go to **62.**
    (c) 81   Go to **63.**

61

Your answer is incorrect; 4^3 is not equal to 12. The raised 3 in 4^3 tells you to multiply 4 by itself. Use the number 4 <u>three</u> times as a factor in a multiplication.

$$4^3 = \underbrace{4 \times 4 \times 4}$$
$$\text{means use three}$$
$$\text{factors of 4}$$

Once you have set up the multiplication in this way it is easy to do it.

$$4 \times 4 \times 4 = (4 \times 4) \times 4 = 16 \times 4 = 64$$

Now, go on to **62.**

62

Excellent! In the exponential form 4^3, the exponent 3 tells you to multiply three factors of 4, that is

$$4^3 = 4 \times 4 \times 4 = (4 \times 4) \times 4 = 16 \times 4 = 64$$

It is important that you be able to read exponential forms correctly.

2^2 is read "two to the second power" or "two squared,"
2^3 is read "two to the third power" or "two cubed,"
2^4 is read "two to the fourth power,"
2^5 is read "two to the fifth power," and so on.

Do the following problems to help get these concepts into your mental muscles. (Reading about a new concept may get it into, or at least through, your head, but only doing problems will make it part of you.)

(a) Write in exponential form:

$5 \times 5 \times 5 \times 5$ = _____ base = _____ exponent = _____

7×7 = _____ base = _____ exponent = _____

$10 \times 10 \times 10 \times 10 \times 10$ = _____ base = _____ exponent = _____

$3 \times 3 \times 3 \times 3 \times 3 \times 3 \times 3$ = _____ base = _____ exponent = _____

$9 \times 9 \times 9$ = _____ base = _____ exponent = _____

$1 \times 1 \times 1 \times 1$ = _____ base = _____ exponent = _____

(b) Write as a product of factors and multiply out:

2^6 = _____ = _____ base = _____ exponent = _____

10^7 = _____ = _____ base = _____ exponent = _____

3^4 = _____ = _____ base = _____ exponent = _____

5^2 = _____ = _____ base = _____ exponent = _____

6^3 = _____ = _____ base = _____ exponent = _____

4^5 = _____ = _____ base = _____ exponent = _____

12^3 = _____ = _____ base = _____ exponent = _____

1^5 = _____ = _____ base = _____ exponent = _____

The correct answers are in **64**. Go there when you finish these.

63

This answer is not correct.   Apparently you found the product $3 \times 3 \times 3 \times 3$.

$$4\overset{3}{\underset{\uparrow}{}} = \underbrace{4 \times 4 \times 4}_{}$$
$$\text{three factors of 4}$$

The raised 3 tells you how many factors of 4 are to be multiplied together.

$$4 \times 4 \times 4 = (4 \times 4) \times 4 = 16 \times 4 = ?$$

Complete this and return to **60** to continue.

64

(a) $5 \times 5 \times 5 \times 5 = 5^4$; base = 5, exponent = 4
$7 \times 7 = 7^2$; base = 7, exponent = 2
$10 \times 10 \times 10 \times 10 \times 10 = 10^5$; base = 10, exponent = 5
$3 \times 3 \times 3 \times 3 \times 3 \times 3 \times 3 = 3^7$; base = 3, exponent = 7
$9 \times 9 \times 9 = 9^3$; base = 9, exponent = 3
$1 \times 1 \times 1 \times 1 = 1^4$; base = 1, exponent = 4

(b) $2^6 = 2 \times 2 \times 2 \times 2 \times 2 \times 2 = 64$; base = 2, exponent = 6
$10^7 = 10 \times 10 \times 10 \times 10 \times 10 \times 10 = 10,000,000$; base = 10,
     exponent = 7
$3^4 = 3 \times 3 \times 3 \times 3 = 81$; base = 3, exponent = 4
$5^2 = 5 \times 5 = 25$; base = 5, exponent = 2
$6^3 = 6 \times 6 \times 6 = 216$; base = 6, exponent = 3
$4^5 = 4 \times 4 \times 4 \times 4 \times 4 = 1024$; base = 4, exponent = 5
$12^3 = 12 \times 12 \times 12 = 1728$; base = 12, exponent = 3
$1^5 = 1 \times 1 \times 1 \times 1 \times 1 = 1$; base = 1, exponent = 5

● Any power of 1 is equal to 1.

$$1^2 = 1 \times 1 = 1$$
$$1^3 = 1 \times 1 \times 1 = 1$$
$$1^4 = 1 \times 1 \times 1 \times 1 = 1, \text{ and so on}$$

Notice that when the base is ten, the product is easy to find.

$$10^2 = 100$$
$$10^3 = 1,000$$
$$10^4 = 10,000$$
$$10^5 = 100,000, \text{ and so on}$$

The exponent number is always exactly equal to the number of zeros in the final product.

Look at that sequence of powers of ten again.   Can you guess the value of 10^1?   What about 10^0?   Can you continue the pattern?

$10^1 = $ _____ $\qquad\qquad$ $10^0 = $ _____

Try it, then turn to **65.**

65

In the pattern

$$10^5 = 100,000$$
$$10^4 = 10,000$$
$$10^3 = 1,000$$
$$10^2 = 100$$

each product on the right decreases by a factor of 10.   Therefore, the next two lines must be

$$10^1 = 10$$
$$10^0 = 1$$

Of course, this is true for any base.

$$2^1 = 2, \ 2^0 = 1;$$
$$3^1 = 3, \ 3^0 = 1;$$
$$4^1 = 4, \ 4^0 = 1; \text{ and so on.}$$

If we factor the number 2592 into its prime factors we find that

$$2592 = 2 \times 2 \times 2 \times 2 \times 2 \times 3 \times 3 \times 3 \times 3$$

Write this using exponents.

$2592 = $ _____

Check your work in **66.**

Why isn't 2° equal to 0?

Remember, we're not multiplying 2×0 in 2°. We define 2° = 1 so that each power of 2 is one factor of 2 larger than the last... 1,2,4,8,16,.......

66

$$2592 = 2^5 \times 3^4$$

Using exponents provides a simple and compact way to write any number as a product of its prime factors.

Find the following products by multiplying:

$2^4 \times 5^3 =$ _____ $3^4 \times 2^7 \times 1^6 =$ _____

$6^2 \times 7^3 =$ _____ $5^3 \times 8^2 \times 2^0 =$ _____

$2^3 \times 3^2 \times 5^4 =$ _____ $7^2 \times 9^1 \times 3^5 =$ _____

$2^2 \times 3^3 \times 4^4 =$ _____ $3^4 \times 5^2 \times 7^1 =$ _____

Go to **67** to check your answers.

Is there any quick way to get 3^8 without multiplying it out the long way?

No. But powers of 10 can be written down with no work! See frame 65.

67

$2^4 \times 5^3 = 2000$ $3^4 \times 2^7 \times 1^6 = 10,368$
$6^2 \times 7^3 = 12,348$ $5^3 \times 8^2 \times 2^0 = 8,000$
$2^3 \times 3^2 \times 5^4 = 45,000$ $7^2 \times 9^1 \times 3^5 = 107,163$
$2^2 \times 3^3 \times 4^4 = 27,648$ $3^4 \times 5^2 \times 7^1 = 14,175$

What is interesting about the following numbers?

1, 4, 9, 16, 25, 36, 49, 64, 81, 100

Do you recognize them?   See **68.**

68

These numbers are the squares or second powers of the counting numbers.

$$1^2 = 1$$
$$2^2 = 4$$
$$3^2 = 9$$
$$4^2 = 16, \text{ and so on.}$$

1, 4, 9, 16, 25 . . . are called <u>perfect</u> <u>squares</u>. If you have memorized the multiplication table for one-digit numbers you will recognize them immediately. If you do not remember them it will be helpful to you to memorize them. A study card for them is provided in the back of this book (page 241). Here are the first twenty perfect squares:

PERFECT SQUARES

$1^2 = 1$	$6^2 = 36$	$11^2 = 121$	$16^2 = 256$
$2^2 = 4$	$7^2 = 49$	$12^2 = 144$	$17^2 = 289$
$3^2 = 9$	$8^2 = 64$	$13^2 = 169$	$18^2 = 324$
$4^2 = 16$	$9^2 = 81$	$14^2 = 196$	$19^2 = 361$
$5^2 = 25$	$10^2 = 100$	$15^2 = 225$	$20^2 = 400$

The number 3^2 is read "three squared." What is "square" about $3^2 =$ 9? The name comes from an old Greek idea about the nature of numbers. Ancient Greek mathematicians called certain numbers "square numbers" or "perfect squares" because they could be represented by a square array of dots.

4   9   16   . . . and so on.

The number of dots along the side of the square was called the "root" or origin of the square number. We call it the square root. For example, the square root of 16 is 4, since $4 \times 4 = 16$.

What is the square root of 64?

(a) 32   Go to **69.**
(b) 8   Go to **70.**

69

Sorry, you are incorrect. We cannot simply divide 64 in half to find its square root!

The square root of 64 is some number ☐ such that ☐ × ☐ = 64. For example, the square root of 25 is equal to 5 because $5 \times 5 = 25$. To "square" means to multiply by itself and to find a square root means to find a number that when multiplied by itself gives the original number.

Now return to **68** and choose a better answer.

70

Right! The square root of 64 is equal to 8 because $8 \times 8 = 64$.
The sign $\sqrt{}$ is used to indicate the square root.

$\sqrt{16} = 4$ is read "square root of 16 equals 4."

$\sqrt{9} = 3$ is read "square root of 9 equals 3."

$\sqrt{169} = 13$ is read "square root of 169 equals 13."

Find these square roots:

$\sqrt{81}$ = _____ $\qquad$ $\sqrt{361}$ = _____ $\qquad$ $\sqrt{289}$ = _____

Try using the table in **68** if you do not recognize these. Continue in **71**.

Where did that funny little $\sqrt{\ }$ sign come from?

The word "root" is <u>radix</u> in Latin and about 1525 A.D. someone began to abbreviate it with the letter r in handwriting. Soon r led to ɼ to √ to √.

71

$\sqrt{81}$ = 9 $\qquad$ Check: $9 \times 9 = 81$

$\sqrt{361}$ = 19 $\qquad$ Check: $19 \times 19 = 361$

$\sqrt{289}$ = 17 $\qquad$ Check: $17 \times 17 = 289$

Always check your answer as shown.

How do you find the square root of any whole number? The surest and simplest way is to consult a table of square roots. There is no easy way to recognize or identify perfect squares. You will find such a table of square roots on page 237 of this book. However, most square roots listed there are not whole numbers. Modern mathematicians have extended the Greek idea of square roots from perfect squares to all numbers.

Now, for a set of practice problems on exponents and square roots, turn to **72**.

72

Problem Set 8: Exponents and Square Roots

A. Find the value of these:

2^4 = _____ $\qquad$ 3^2 = _____ $\qquad$ 4^3 = _____ $\qquad$ 5^3 = _____

10^3 = _____ $\qquad$ 7^2 = _____ $\qquad$ 2^8 = _____ $\qquad$ 6^2 = _____

8^3 = _____ $\qquad$ 3^4 = _____ $\qquad$ 5^4 = _____ $\qquad$ 10^5 = _____

$2^3 =$ _____ $3^5 =$ _____ $9^3 =$ _____ $7^0 =$ _____

$6^1 =$ _____ $1^4 =$ _____ $4^4 =$ _____ $2^5 =$ _____

$10^6 =$ _____ $7^3 =$ _____ $8^2 =$ _____ $6^4 =$ _____

$2^{10} =$ _____ $9^4 =$ _____ $6^3 =$ _____ $5^2 =$ _____

$3^3 =$ _____ $7^4 =$ _____ $9^0 =$ _____ $10^4 =$ _____

$4^2 =$ _____ $5^1 =$ _____ $8^4 =$ _____ $1^{14} =$ _____

$9^2 =$ _____ $4^5 =$ _____ $10^2 =$ _____ $6^5 =$ _____

B. Find the value of these:

$14^2 =$ _____ $21^2 =$ _____ $15^3 =$ _____

$25^3 =$ _____ $16^2 =$ _____ $55^2 =$ _____

$61^2 =$ _____ $40^3 =$ _____ $100^3 =$ _____

$2^2 \times 3^3 =$ _____ $2^6 \times 3^2 =$ _____ $3^2 \times 5^3 =$ _____

$2^1 \times 3^4 \times 7^2 =$ _____ $3^2 \times 7^2 \times 11^1 =$ _____ $2^0 \times 3^4 \times 5^2 =$ _____

$2^3 \times 7^3 =$ _____ $9^2 \times 2^4 =$ _____ $6^2 \times 5^2 \times 3^3 =$ _____

$2^1 \times 10^3 =$ _____ $3^2 \times 10^4 =$ _____ $2^{10} \times 3^2 =$ _____

C. Calculate these square roots:

$\sqrt{81} =$ _____ $\sqrt{144} =$ _____ $\sqrt{16} =$ _____ $\sqrt{25} =$ _____

$\sqrt{36} =$ _____ $\sqrt{100} =$ _____ $\sqrt{49} =$ _____ $\sqrt{324} =$ _____

$\sqrt{1} =$ _____ $\sqrt{121} =$ _____ $\sqrt{64} =$ _____ $\sqrt{9} =$ _____

$\sqrt{225} =$ _____ $\sqrt{4} =$ _____ $\sqrt{400} =$ _____ $\sqrt{256} =$ _____

D. Brain Boosters

1. Show that the following interesting equations are correct.

$2^5 \times 9^2 = 2592$ $3^4 \times 425 = 34,425$

$12^2 + 33^2 = 1233$ $31^2 \times 325 = 312,325$

$88^2 + 33^2 = 8833$ $87^2 - 78^2 = 41^2 - 14^2$

$1^3 + 5^3 + 3^3 = 153$ $75^2 - 57^2 = 51^2 - 15^2$

$3^3 + 7^3 + 1^3 = 371$ $9^4 + 4^4 + 7^4 + 4^4 = 9474$

$4^3 + 0^3 + 7^3 = 407$ $3^3 + 4^4 + 3^3 + 5^5 = 3435$

$1^1 + 3^2 + 5^3 = 135$ $(4 + 9 + 1 + 3)^3 = 4913$ (Add the four

$1^1 + 7^2 + 5^3 = 175$ numbers in the parentheses, then
                                          cube their sum.)

$$4^2 + 3^3 = 43 \qquad\qquad 6^2 + 3^3 = 63$$

(Can you see why they are interesting?)

2. Show that this set of equations is true by finding the value of each side.

$$2 + 3 + 10 + 11 = 1 + 5 + 8 + 12$$
$$2^2 + 3^2 + 10^2 + 11^2 = 1^2 + 5^2 + 8^2 + 12^2$$
$$2^3 + 3^3 + 10^3 + 11^3 = 1^3 + 5^3 + 8^3 + 12^3$$

3. Check the following claims for the first few powers.

    (a) Every power of 5 ends in 5.

    $5^2 = $ _____ $5^3 = $ _____ $5^4 = $ _____

    (b) Every power of 25 ends in 25.

    $25^2 = $ _____ $25^3 = $ _____ $25^4 = $ _____

    (c) Every power of 625 ends in 625.

    $625^2 = $ _____ $625^3 = $ _____ $625^4 = $ _____

    (d) Every power of 376 ends in 376.

    $376^2 = $ _____ $376^3 = $ _____ $376^4 = $ _____

The answers to these problems are on pages 225–226. When you have had the practice you need turn to **73** for a self-test covering the concepts learned in Chapter 1.

Chapter 1 Self-Test

1. $37 + 46 =$ _____

2. $4372 + 849 =$ _____

3. $81 + 47 + 36 =$ _____

4. $43 - 17 =$ _____

5. $702 - 416 =$ _____

6. $6001 - 3973 =$ _____

7. $8300 - 4605 =$ _____

8. $43 \times 36 =$ _____

9. $237 \times 204 =$ _____

10. $406 \times 137 =$ _____

11. $2153 \times 302 =$ _____

12. $1081 \div 47 =$ _____

13. $27\overline{)21687} =$ _____

14. $2008 \div 4 =$ _____

15. $\dfrac{26 \times 48}{39} =$ _____

16. Factor: $408 =$ _____

17. Factor: $9438 =$ _____

18. Factor: $1000 =$ _____

19. Factor: $3264 =$ _____

20. $3^4 =$ _____

21. $3^0 \times 4^2 \times 5^3 =$ _____

22. $2^1 \times 3^4 \times 7^2 =$ _____

23. $123^2 =$ _____

24. $\sqrt{225} =$ _____

25. $\sqrt{324} =$ _____

The answers to these problems are on page 234.

CHAPTER TWO

Fractions

Preview

Objectives		

1. Rename fractions:
   write a fraction equivalent to any given fraction;

   (a) $\dfrac{3}{4} = \dfrac{?}{12}$

   reduce a fraction to lowest terms;

   (b) $\dfrac{18}{39} = \underline{\hspace{2cm}}$

   write a mixed number as a fraction;

   (c) $4\dfrac{2}{3} = \underline{\hspace{2cm}}$

   compare fractions.

   (d) Which is larger: $\dfrac{3}{13}$ or $\dfrac{4}{17}$?

2. Multiply and divide fractions.

   (a) $1\dfrac{1}{3} \times 2\dfrac{3}{5} = \underline{\hspace{2cm}}$

   (b) $\left(2\dfrac{2}{3}\right)^2 = \underline{\hspace{2cm}}$

   (c) $3\dfrac{1}{2} \div 1\dfrac{3}{4} = \underline{\hspace{2cm}}$

	Page	Frame
(d) $6 \div 2\frac{1}{3} =$ _____	98	**26**
(e) Divide 6 by $2\frac{2}{5}$. _____	98	**26**

3. Add and subtract fractions.

	Page	Frame
(a) $4\frac{2}{3} + 1\frac{3}{4} =$ _____	107	**34**
(b) $5\frac{1}{8} - 3\frac{1}{3} =$ _____	107	**34**
(c) $9 - 1\frac{3}{8} =$ _____	107	**34**

4. Do word problems involving fractions.

	Page	Frame
(a) What fraction of $\frac{2}{3}$ is $1\frac{3}{4}$? _____	120	**51**
(b) Find $2\frac{1}{3}$ of $1\frac{7}{8}$. _____	120	**51**
(c) If 7 apples cost 91¢, what will 4 apples cost? _____	120	**51**
(d) Find a number such that $\frac{7}{8}$ of it is $2\frac{1}{2}$. _____	120	**51**

If you are certain you can work all of these problems correctly, turn to page 133 for a self-test. If you want help with any of these objectives or if you cannot work one of the preview problems, turn to the page indicated. Super-students (those who want to be certain they learn all of this) will join us in frame **1**.

1. (a) 9

   (b) $\dfrac{6}{13}$

   (c) $\dfrac{14}{3}$

   (d) $\dfrac{4}{17}$

2. (a) $\dfrac{52}{15}$ or $3\dfrac{7}{15}$

   (b) $\dfrac{64}{9}$ or $7\dfrac{1}{9}$

   (c) 2

   (d) $\dfrac{18}{7}$ or $2\dfrac{4}{7}$

   (e) $\dfrac{5}{2}$ or $2\dfrac{1}{2}$

3. (a) $\dfrac{77}{12}$ or $6\dfrac{5}{12}$

   (b) $\dfrac{43}{24}$ or $1\dfrac{19}{24}$

   (c) $\dfrac{61}{8}$ or $7\dfrac{5}{8}$

4. (a) $\dfrac{21}{8}$ or $2\dfrac{5}{8}$

   (b) $\dfrac{35}{8}$ or $4\dfrac{3}{8}$

   (c) 52¢

   (d) $\dfrac{20}{7} = 2\dfrac{6}{7}$

ANSWERS TO PREVIEW PROBLEMS

Renaming Fractions

1

By permission of Johnny Hart and Field Enterprises Inc.

Every caveman knows that measuring sticks must have numbers to label their marks, but only very smart ones like our friend in the cartoon learn to talk about the space between markers. We use counting numbers to count measurement units: 1 inch, 2 inches, 3 inches, or even 4 centimeters if you are a very modern caveman. Fractions enable us to label some of the points between counting numbers. The word fraction comes from the Latin fractus meaning to break. We use fractions to describe subdivisions of the standard measurement units for length, time, money, or whatever we choose to measure.

Consider the rectangular area below. What happens when we break it into equal parts?

2 parts, each one-half of the whole

3 parts, each one-third of the whole

4 parts, each one-fourth of the whole

Divide the area below into fifths by drawing vertical lines.

Try it, then go to **2.**

2

Notice that the five parts or "fifths" are equal in area.

A fraction is normally written as the division of two whole numbers:

$$\frac{2}{3}, \frac{3}{4}, \text{ or } \frac{26}{12}$$

Each of the five equal areas above would be "one-fifth" or $\frac{1}{5}$ of the entire area.

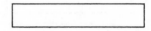 $\frac{1}{5} = \frac{1 \text{ shaded part}}{5 \text{ parts total}}$

How would you label the shaded portion of the area below?

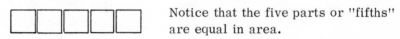

Continue in **3.**

3

 $\frac{3}{5} = \frac{3 \text{ shaded parts}}{5 \text{ parts total}}$

The fraction $\frac{3}{5}$ implies an area equal to three of the original portions.

$$\frac{3}{5} = 3 \times \left(\frac{1}{5}\right)$$

There are three equal parts and the name of each part is $\frac{1}{5}$ or one-fifth.

In the collection of letters below, what fraction are **H**s? (Hint: Count the total number of letters and decide what portion are **H**s.)

H H H H S S T

Check your answer in **4**.

4

$$\text{fraction of } \mathbf{H}\text{s} = \frac{\text{number of } \mathbf{H}\text{s}}{\text{total number of letters}} = \frac{4}{7}$$

The answer, $\frac{4}{7}$, is read "four-sevenths." The fraction of **S**s is $\frac{2}{7}$ and the fraction of **T**s is $\frac{1}{7}$.

The two numbers that form a fraction are given special names to simplify talking about them. In the fraction $\frac{3}{5}$ the upper number (3) is called the <u>numerator</u> from the Latin <u>numero</u> meaning number. It is a count of the number of parts. The lower number (5) is called the <u>denominator</u> from the Latin <u>nomen</u> or <u>name</u>. It tells us the name of the part being counted.

$\underline{3}$ ← numerator, the number of parts
5 ← denominator, the name of the part ("fifths," in this case)

A textbook costs \$6 and I have \$5. What fraction of its cost do I have? Write the answer as a fraction.

_____ numerator = _____ , denominator = _____

Check your answer in **5**.

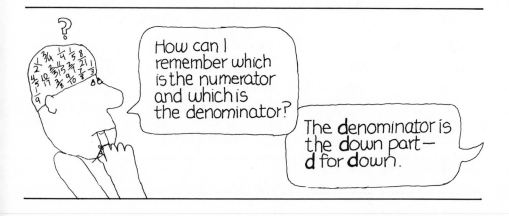

5

$$\underbrace{\$\,\$\,\$\,\$\,\$}_{5}$$

$5 is $\dfrac{5}{6}$ of the total cost.

numerator = 5, denominator = 6

Complete these sentences by writing in the correct fraction.

(a) If we divide a length into eight equal parts, each part will be _____ of the total length.

(b) Then three of these parts will represent _____ of the total length.

(c) Eight of these parts will be _____ of the total length.

(d) Ten of these parts will be _____ of the total length.

Check your answers in **6.**

6

(a) $\dfrac{1}{8}$; (b) $\dfrac{3}{8}$; (c) $\dfrac{8}{8}$; (d) $\dfrac{10}{8}$

The original length is used as a standard and any other length (smaller or larger) can be expressed as a fraction of the original length. A <u>proper fraction</u> is a number less than 1, as you would suppose a fraction should be. It represents a quantity less than the standard. For example,

$$\frac{1}{2}, \frac{2}{3}, \text{ and } \frac{17}{20}$$

are all proper fractions. Notice that for a proper fraction, the numerator is less than the denominator (the top number is less than the bottom number).

An <u>improper fraction</u> is a number greater than 1 and represents a quantity greater than the standard. If a standard length is 8 inches, a length of 11 inches would be $\dfrac{11}{8}$ of the standard. Notice that for an improper fraction the numerator is greater than the denominator (top number greater than the bottom number).

Circle the proper fractions in the following list.

$$\frac{3}{2} \quad \frac{3}{4} \quad \frac{7}{8} \quad \frac{5}{4} \quad \frac{15}{12} \quad \frac{1}{16} \quad \frac{35}{32} \quad \frac{7}{50} \quad \frac{65}{64} \quad \frac{105}{100}$$

Go to **7** when you have finished.

7

You should have circled the following proper fractions: $\frac{3}{4}, \frac{7}{8}, \frac{1}{16}, \frac{7}{50}$.

All these are numbers less than 1. In each the numerator is less than the denominator.

The improper fraction $\frac{7}{3}$ can be shown graphically as follows:

unit standard =

$$\frac{1}{3} = $$

then, $\frac{7}{3} = $ (seven, count 'em)

We can rename this number by regrouping.

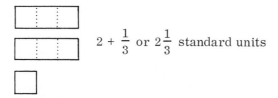

$2 + \frac{1}{3}$ or $2\frac{1}{3}$ standard units

A <u>mixed</u> <u>number</u> is an improper fraction written as the sum of a whole number and a proper fraction.

$$\frac{7}{3} = 2 + \frac{1}{3} \text{ or } 2\frac{1}{3}$$

We usually omit the plus sign and write $2 + \frac{1}{3}$ as $2\frac{1}{3}$, and read it as "two and one-third." The numbers $1\frac{1}{2}$, $2\frac{3}{5}$, and $16\frac{2}{3}$ are all written as mixed numbers.

To write an improper fraction as a mixed number, divide numerator by denominator and form a new fraction as shown on the next page.

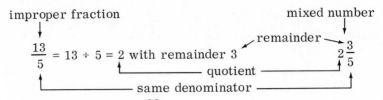

improper fraction · mixed number

$$\frac{13}{5} = 13 \div 5 = 2 \text{ with remainder } 3 \qquad 2\frac{3}{5}$$

remainder · quotient · same denominator

Now you try it.   Rename $\frac{23}{4}$ as a mixed number.

$$\frac{23}{4} = \underline{\hspace{2cm}}$$

Follow the procedure shown above, then turn to **9.**

8

(a) $\frac{9}{5} = 1\frac{4}{5}$ (b) $\frac{13}{4} = 3\frac{1}{4}$ (c) $\frac{27}{8} = 3\frac{3}{8}$

(d) $\frac{31}{5} = 6\frac{1}{5}$ (e) $\frac{41}{12} = 3\frac{5}{12}$ (f) $\frac{17}{2} = 8\frac{1}{2}$

The reverse process, rewriting a mixed number as an improper fraction, is equally simple.   Study this example of $2\frac{3}{5}$ being converted to an improper fraction.

■Work in a clockwise direction.   First multiply $5 \times 2 = 10$.

$$2 \overset{\times}{\frown} \frac{3}{5}$$

■Then add the numerator, $10 + 3 = 13$.

$$2 \overset{+}{\underset{\times}{\frown}} \frac{3}{5} = \overset{10+3}{\frac{13}{5}}$$

same denominator

Here is how the same operation looks using a picture:

$$2\frac{3}{5} = \begin{cases} \quad \frac{5}{5} \\ \quad \frac{5}{5} \\ \quad \frac{3}{5} \end{cases} \qquad \frac{5}{5} + \frac{5}{5} + \frac{3}{5} = \frac{13}{5}$$

(count 'em)

Now you try it. Rewrite these mixed numbers as improper fractions.

(a) $3\frac{1}{6}$ (b) $4\frac{3}{5}$ (c) $1\frac{1}{2}$

(d) $8\frac{2}{3}$ (e) $15\frac{3}{8}$ (f) $9\frac{3}{4}$

Check your answers in **10.**

9

$$\frac{23}{4} = 23 \div 4 = 5 \text{ with remainder } 3 \longrightarrow 5\frac{3}{4}$$

If in doubt, check your work with a diagram like this:

$$\bullet = \begin{matrix}\bullet\bullet\bullet\bullet\end{matrix}$$

23   5 rows of 4

3 remaining

Now try these for practice. Write each improper fraction as a mixed number.

(a) $\frac{9}{5}$ (b) $\frac{13}{4}$ (c) $\frac{27}{8}$

(d) $\frac{31}{5}$ (e) $\frac{41}{12}$ (f) $\frac{17}{2}$

The answers are in **8.**

10

(a) $3\frac{1}{6} = \frac{19}{6}$ (b) $4\frac{3}{5} = \frac{23}{5}$ (c) $1\frac{1}{2} = \frac{3}{2}$

(d) $8\frac{2}{3} = \frac{26}{3}$ (e) $15\frac{3}{8} = \frac{123}{8}$ (f) $9\frac{3}{4} = \frac{39}{4}$

Two fractions are said to be <u>equivalent</u> if they are numerals or names for the same number. For example,

$$\frac{1}{2} = \frac{2}{4}$$

since both fractions represent the same portion of some standard amount.

$\frac{1}{2}$ $= \frac{2}{4}$

There is a very large set of fractions equivalent to $\frac{1}{2}$:

$$\frac{1}{2} = \frac{2}{4} = \frac{3}{6} = \frac{4}{8} = \frac{5}{10} = \cdots = \frac{46}{92} = \frac{61}{122} = \frac{1437}{2874} \cdots$$

Each fraction is a name for the same number and we can use these fractions interchangeably.

To obtain a fraction equivalent to any given fraction multiply the original (numerator and denominator) by the same nonzero number. For example,

$$\frac{1}{2} = \frac{1 \times 3}{2 \times 3} = \frac{3}{6}$$

$$\frac{2}{3} = \frac{2 \times 5}{3 \times 5} = \frac{10}{15}$$

Rename this fraction as shown:

$$\frac{3}{4} = \frac{?}{20}$$

Check your work in **11.**

11

$$\frac{3}{4} = \frac{3 \times ?}{4 \times ?} = \frac{3 \times 5}{4 \times 5} = \frac{15}{20} \qquad 4 \times ? = 20 \text{ so } ? \text{ must be 5.}$$

The number value of the fraction has not changed; we have simply renamed it.

Practice with these.

(a) $\dfrac{5}{6} = \dfrac{?}{42}$ 　　　　　　(b) $\dfrac{7}{16} = \dfrac{?}{48}$

(c) $\dfrac{3}{7} = \dfrac{?}{56}$ 　　　　　　(d) $1\dfrac{2}{3} = \dfrac{?}{12}$

Look in **13** for the answers.

12

$$\frac{90}{105} = \frac{2 \times \cancel{3} \times 3 \times \cancel{5}}{\cancel{3} \times \cancel{5} \times 7} = \frac{2 \times 3}{7} = \frac{6}{7}$$

This process of eliminating common factors is usually called <u>cancelling</u>. When you cancel a factor you <u>divide</u> both top and bottom of the fraction by that factor. In the cancellation above we actually divided top and bottom by 3×5 or 15.

$$\frac{90}{105} = \frac{90 \div 15}{105 \div 15} = \frac{6}{7}$$

Be Careful

● Cancel the same factor.

● Cancel only multiplying factors.

This is illegal: $\dfrac{2 \times \cancel{3} \times 5}{\cancel{3} \times \cancel{3} \times 7}$

Cancel only one 3. We divide the top by 3, so we must divide the bottom by 3.

This is illegal: $\dfrac{2 + \cancel{5}}{3 + \cancel{5}}$

5 is not a multiplier. Cancelling means dividing the top and bottom of the fraction by the same number. It is illegal to subtract a number from top and bottom.

Reduce the following fractions to lowest terms:

(a) $\dfrac{15}{84}$

(b) $\dfrac{21}{35}$

(c) $\dfrac{4}{12}$

(d) $\dfrac{154}{1078}$

(e) $\dfrac{256}{208}$

(f) $\dfrac{378}{405}$

The answers are in **14**.

13

(a) $\dfrac{5}{6} = \dfrac{5 \times 7}{6 \times 7} = \dfrac{35}{42}$

(b) $\dfrac{7}{16} = \dfrac{7 \times 3}{16 \times 3} = \dfrac{21}{48}$

(c) $\dfrac{3}{7} = \dfrac{3 \times 8}{7 \times 8} = \dfrac{24}{56}$

(d) $1\dfrac{2}{3} = \dfrac{5}{3} = \dfrac{5 \times 4}{3 \times 4} = \dfrac{20}{12}$

Very often in working with fractions you will be asked to <u>reduce a</u> <u>fraction to lowest terms</u>. This means to replace the fraction with the simplest fraction in its set of equivalent fractions. For example, to reduce $\dfrac{15}{30}$ to its lowest terms you would replace it with $\dfrac{1}{2}$.

$$\dfrac{15}{30} = \dfrac{1 \times 15}{2 \times 15} = \dfrac{1}{2}$$

The two fractions $\dfrac{15}{30}$ and $\dfrac{1}{2}$ are equivalent and $\dfrac{1}{2}$ is the simplest equivalent fraction to $\dfrac{15}{30}$ because its numerator (1) and denominator (2) are the smallest whole numbers of any in the set

$$\dfrac{1}{2}, \dfrac{2}{4}, \dfrac{3}{6}, \dfrac{4}{8} \cdots \dfrac{15}{30} \cdots$$

How can you find the simplest equivalent fraction? For example, how would you reduce $\dfrac{30}{42}$ to lowest terms? First, factor numerator and denominator.

$$\dfrac{30}{42} = \dfrac{2 \times 3 \times 5}{2 \times 3 \times 7}$$

Second, identify and eliminate common factors.

$$\frac{30}{42} = \frac{2 \times 3 \times 5}{2 \times 3 \times 7}$$

2 is a common factor, cancel the 2s: $\dfrac{30}{42} = \dfrac{\cancel{2} \times 3 \times 5}{\cancel{2} \times 3 \times 7}$

3 is a common factor, cancel the 3s: $\dfrac{30}{42} = \dfrac{\cancel{2} \times \cancel{3} \times 5}{\cancel{2} \times \cancel{3} \times 7}$

$$\frac{30}{42} = \frac{5}{7}$$

In effect, we have divided both top and bottom of the fraction by $2 \times 3 = 6$, the common factor.

$$\frac{30}{42} = \frac{30 \div 6}{42 \div 6} = \frac{5}{7}$$

Your turn. Reduce $\dfrac{90}{105}$ to lowest terms. Look in **12** for the answer.

14

(a) $\dfrac{15}{84} = \dfrac{\cancel{3} \times 5}{2 \times 2 \times \cancel{3} \times 7} = \dfrac{5}{28}$ (b) $\dfrac{21}{35} = \dfrac{3 \times \cancel{7}}{5 \times \cancel{7}} = \dfrac{3}{5}$

(c) $\dfrac{4}{12} = \dfrac{\cancel{4}}{\cancel{4} \times 3} = \dfrac{1}{3}$ (d) $\dfrac{154}{1078} = \dfrac{\cancel{2} \times \cancel{7} \times \cancel{11}}{\cancel{2} \times \cancel{7} \times 7 \times \cancel{11}} = \dfrac{1}{7}$

(e) $\dfrac{256}{208} = \dfrac{\cancel{2} \times \cancel{2} \times \cancel{2} \times \cancel{2} \times 2 \times 2 \times 2 \times 2}{\cancel{2} \times \cancel{2} \times \cancel{2} \times \cancel{2} \times 13} = \dfrac{16}{13}$

(f) $\dfrac{378}{405} = \dfrac{2 \times \cancel{3} \times \cancel{3} \times \cancel{3} \times 7}{\cancel{3} \times \cancel{3} \times \cancel{3} \times 3 \times 5} = \dfrac{14}{15}$

▷ Remember: Cancellation is division by a common factor.

Reduce the fraction $\dfrac{6}{3}$ to lowest terms. Check your answer in **15**.

15

$$\frac{6}{3} = \frac{2 \times \cancel{3}}{1 \times \cancel{3}} = \frac{2}{1} \text{ (or simply 2)}$$

Any whole number may be written as a fraction by using a denominator equal to 1.

$$3 = \frac{3}{1}, \quad 4 = \frac{4}{1}, \quad \text{and so on}$$

The number 1 can be written as any fraction whose numerator and denominator are equal.

$$1 = \frac{2}{2} = \frac{3}{3} = \frac{4}{4} = \ldots = \frac{72}{72} = \frac{1257}{1257} \ldots \text{ and so on}$$

If you were offered your choice between $\frac{2}{3}$ of a certain amount of money and $\frac{5}{8}$ of the same amount of money, which would you choose? Which is the larger fraction, $\frac{2}{3}$ or $\frac{5}{8}$? Can you decide? Try. Renaming the fractions would help. The answer is in **16**.

16

To compare two fractions rename each by changing them to equivalent fractions with the same denominator.

$$\frac{2}{3} = \frac{2 \times 8}{3 \times 8} = \frac{16}{24} \qquad\qquad \frac{5}{8} = \frac{5 \times 3}{8 \times 3} = \frac{15}{24}$$

Now compare the new fractions: $\frac{16}{24}$ is greater than $\frac{15}{24}$.

▷ Notice: 1. The new denominator is the product of the original denominators ($24 = 8 \times 3$).

2. Once both fractions are written with the same denominator, the one with the larger numerator is the larger. (16 of the fractional parts is more than 15 of them.)

Which of the following pairs of fractions is larger?

(a) $\dfrac{3}{4}$ and $\dfrac{5}{7}$

(b) $\dfrac{7}{8}$ and $\dfrac{19}{21}$

(c) 3 and $\dfrac{40}{13}$

(d) $1\dfrac{7}{8}$ and $\dfrac{5}{3}$

(e) $2\dfrac{1}{4}$ and $\dfrac{11}{5}$

(f) $\dfrac{5}{16}$ and $\dfrac{11}{35}$

Check your answers in **17.**

17

(a) $\dfrac{3}{4} = \dfrac{21}{28}$, $\dfrac{5}{7} = \dfrac{20}{28}$; $\dfrac{21}{28}$ is larger than $\dfrac{20}{28}$ so $\dfrac{3}{4}$ is larger than $\dfrac{5}{7}$

(b) $\dfrac{7}{8} = \dfrac{147}{168}$, $\dfrac{19}{21} = \dfrac{152}{168}$; $\dfrac{152}{168}$ is larger than $\dfrac{147}{168}$ so $\dfrac{19}{21}$ is larger than $\dfrac{7}{8}$

(c) $3 = \dfrac{39}{13}$; $\dfrac{40}{13}$ is larger than $\dfrac{39}{13}$ so $\dfrac{40}{13}$ is larger than 3

(d) $1\dfrac{7}{8} = \dfrac{15}{8} = \dfrac{45}{24}$, $\dfrac{5}{3} = \dfrac{40}{24}$; $\dfrac{45}{24}$ is larger than $\dfrac{40}{24}$ so $1\dfrac{7}{8}$ is larger than $\dfrac{5}{3}$

(e) $2\dfrac{1}{4} = \dfrac{9}{4} = \dfrac{45}{20}$, $\dfrac{11}{5} = \dfrac{44}{20}$; $\dfrac{45}{20}$ is larger than $\dfrac{44}{20}$ so $2\dfrac{1}{4}$ is larger than $\dfrac{11}{5}$

(f) $\dfrac{5}{16} = \dfrac{175}{560}$, $\dfrac{11}{35} = \dfrac{176}{560}$; $\dfrac{176}{560}$ is larger than $\dfrac{175}{560}$ so $\dfrac{11}{35}$ is larger than $\dfrac{5}{16}$

Now turn to **18** for some practice renaming fractions.

18

Problem Set 1: Renaming Fractions

A. Write as an improper fraction:

$2\dfrac{1}{3}$ $\qquad$ $4\dfrac{2}{5}$ $\qquad$ $7\dfrac{1}{2}$ $\qquad$ $13\dfrac{3}{7}$ $\qquad$ $8\dfrac{3}{4}$

4 $\qquad$ $1\dfrac{2}{3}$ $\qquad$ $5\dfrac{5}{6}$ $\qquad$ $3\dfrac{7}{8}$ $\qquad$ $2\dfrac{3}{5}$

$16\dfrac{1}{10}$ $\qquad$ $70\dfrac{5}{9}$ $\qquad$ $12\dfrac{1}{40}$ $\qquad$ $15\dfrac{5}{11}$ $\qquad$ $37\dfrac{2}{3}$

B. Write as a mixed number:

| $\dfrac{17}{2}$ | $\dfrac{23}{3}$ | $\dfrac{8}{5}$ | $\dfrac{19}{4}$ | $\dfrac{37}{6}$ | $\dfrac{28}{3}$ | $\dfrac{37}{8}$ | $\dfrac{29}{7}$ |

| $\dfrac{34}{25}$ | $\dfrac{47}{9}$ | $\dfrac{211}{4}$ | $\dfrac{170}{23}$ | $\dfrac{43}{10}$ | $\dfrac{125}{6}$ | $\dfrac{139}{15}$ |

C. Reduce to lowest terms:

| $\dfrac{26}{30}$ | $\dfrac{12}{15}$ | $\dfrac{8}{10}$ | $\dfrac{27}{54}$ | $\dfrac{5}{40}$ |

| $\dfrac{18}{45}$ | $\dfrac{7}{42}$ | $\dfrac{16}{18}$ | $\dfrac{9}{27}$ | $\dfrac{21}{56}$ |

| $\dfrac{42}{120}$ | $\dfrac{54}{144}$ | $\dfrac{36}{216}$ | $\dfrac{280}{490}$ | $\dfrac{115}{207}$ |

D. Complete these:

| $\dfrac{7}{8} = \dfrac{?}{16}$ | $\dfrac{3}{5} = \dfrac{?}{45}$ | $\dfrac{3}{4} = \dfrac{?}{12}$ | $2\dfrac{5}{12} = \dfrac{?}{60}$ | $\dfrac{1}{9} = \dfrac{?}{63}$ |

| $1\dfrac{2}{7} = \dfrac{?}{35}$ | $\dfrac{5}{8} = \dfrac{?}{32}$ | $5\dfrac{3}{5} = \dfrac{?}{25}$ | $\dfrac{1}{2} = \dfrac{?}{78}$ | $\dfrac{2}{3} = \dfrac{?}{51}$ |

| $8\dfrac{1}{4} = \dfrac{?}{44}$ | $5\dfrac{6}{7} = \dfrac{?}{14}$ | $\dfrac{11}{12} = \dfrac{?}{72}$ | $3\dfrac{7}{10} = \dfrac{?}{50}$ | $9\dfrac{5}{9} = \dfrac{?}{54}$ |

E. Brain Boosters

1. A group of runners want to race 1 kilometer, a distance roughly equal to $\dfrac{62}{100}$ of a mile. Which is closer to 1 kilometer—6 laps on a track where each lap is $\dfrac{1}{11}$ of a mile or 5 laps on a track where each lap is $\dfrac{1}{8}$ of a mile?

2. Denny Dimwit wrote the following on his arithmetic exam:

$$\dfrac{1\!\!\!/6}{8\!\!\!/4} = \dfrac{1}{4} \qquad \dfrac{2\!\!\!/6}{6\!\!\!/5} = \dfrac{2}{5}$$

He claims that he has discovered that 6s always cancel. His teacher says that it is accidental that these work. Write Denny a short explanation of what cancelling really means.

3. A box of Sugar Glops breakfast cereal contains $\dfrac{7}{10}$ of a pound of cereal. A box of Astro Puffs contains 11 ounces of cereal. (1 ounce = $\dfrac{1}{16}$ of a pound) The two cereals have the same food value

(almost none!) and cost exactly the same amount. Which is the better buy?

4. Nurse Hypo was supposed to give her patient four $\frac{1}{7}$ grain No-Go pills. Instead, she gave him nine $\frac{1}{16}$ grain pills. Did she give him too much or too little?

5. Each of the following mixed numbers contains all nine nonzero digits and they are all equal to the same whole number. What whole number are they equal to?

(a) $91\frac{7524}{836}$ (b) $91\frac{5823}{647}$ (c) $94\frac{1578}{263}$

(d) $96\frac{2148}{537}$ (e) $96\frac{1428}{357}$ (f) $96\frac{1752}{438}$

The answers to these problems are on page 227. When you have had the practice you need, either return to the preview test on page 76 or continue in **19** with the study of multiplication of fractions.

Multiplication

19

The simplest arithmetic operation with fractions is multiplication and, happily, it is easy to show graphically. The multiplication of a whole number and a fraction may be illustrated this way:

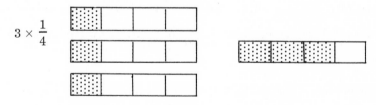

$3 \times \frac{1}{4}$

$3 \times \frac{1}{4} = \frac{1}{4} + \frac{1}{4} + \frac{1}{4} = \frac{3}{4}$ (three segments each $\frac{1}{4}$ unit long)

Any fraction such as $\frac{3}{4}$ can be thought of as a product: $3 \times \frac{1}{4}$.

The product of two fractions can also be shown graphically.

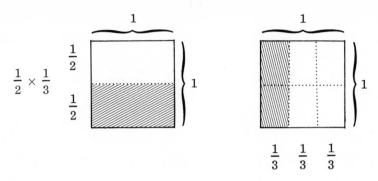

$$\frac{1}{2} \times \frac{1}{3}$$

The product is: $\frac{1}{2} \times \frac{1}{3} = \frac{1}{6} = \dfrac{\text{1 shaded area}}{\text{6 equal areas in the } 1 \times 1 \text{ square}}$

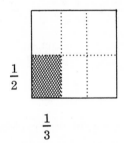

Another way to solve this is $\frac{1}{2} \times \frac{1}{3}$ means $\frac{1}{2}$ of $\frac{1}{3}$.

the whole   ($\frac{1}{3}$ of the whole   ($\frac{1}{2}$ of $\frac{1}{3}$ of the whole or $\frac{1}{6}$ of the whole

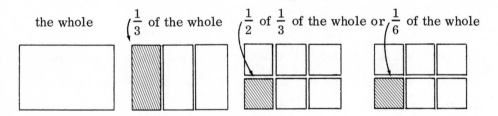

In general, we calculate this product as

$$\frac{1}{2} \times \frac{1}{3} = \frac{1 \times 1}{2 \times 3} = \frac{1}{6}$$

The product of two fractions is a fraction whose numerator is the product of their numerators and whose denominator is the product of their denominators.

Multiply $\frac{5}{6} \times \frac{2}{3}$, then choose which answer below you think is correct.

(a) $\frac{5}{6} \times \frac{2}{3} = \frac{10}{18}$ Go to **20.**

(b) $\dfrac{5}{6} \times \dfrac{2}{3} = \dfrac{5}{9}$ Go to **21.**

(c) I don't know how to do it and I can't figure
out how to draw the little boxes. Go to **22.**

20

Right. $\dfrac{5}{6} \times \dfrac{2}{3} = \dfrac{5 \times 2}{6 \times 3} = \dfrac{10}{18}$

Now reduce this answer to lowest terms and go to **21.**

21

Excellent. $\dfrac{5}{6} \times \dfrac{2}{3} = \dfrac{5 \times 2}{6 \times 3} = \dfrac{5 \times \cancel{2}}{3 \times \cancel{2} \times 3} = \dfrac{5}{9}$

Always reduce your answer to lowest terms. In this problem you probably recognized that 6 was evenly divisible by 2 and did it like this:

$$\dfrac{5}{\cancel{6}_3} \times \dfrac{\cancel{2}^{1}}{3} = \dfrac{5}{9}$$

You divided the top and the bottom of the fraction by 2. It will save you
time and effort if you eliminate common factors (such as the 2 above)
<u>before</u> you multiply.

Do you see that $3 \times \dfrac{1}{4}$ is really the same sort of problem?

$$3 = \dfrac{3}{1}, \text{ so } 3 \times \dfrac{1}{4} = \dfrac{3}{1} \times \dfrac{1}{4} = \dfrac{3 \times 1}{1 \times 4} = \dfrac{3}{4}$$

Test your understanding with these problems. (Hint: Change mixed
numbers such as $1\dfrac{1}{2}$ and $3\dfrac{5}{6}$ into improper fractions; then multiply as
usual.)

(a) $\dfrac{7}{8} \times \dfrac{2}{3} =$ _____

(b) $\dfrac{8}{12} \times \dfrac{3}{16} =$ _____

(c) $\dfrac{3}{32} \times \dfrac{4}{15} =$ _____

(d) $\dfrac{15}{4} \times \dfrac{9}{10} =$ _____

(e) $\dfrac{3}{2} \times \dfrac{2}{3} =$ _____

(f) $1\dfrac{1}{2} \times \dfrac{2}{5} =$ _____

(g) $4 \times \dfrac{7}{8} =$ _____

(h) $3\dfrac{5}{6} \times \dfrac{3}{10} =$ _____

(i) $\dfrac{4}{3} \times \dfrac{3}{4} =$ _____

Have you reduced all answers to lowest terms? The correct answers are in **23**.

22

Don't panic. You don't need to draw the little boxes to do the calculation. Try it this way:

$$\frac{5}{6} \times \frac{2}{3} = \frac{5 \times 2}{6 \times 3} \longleftarrow \text{multiply the numerators} \\ \longleftarrow \text{multiply the denominators}$$

Finish the calculation and then return to **19** and choose an answer.

23

(a) $\dfrac{7}{8} \times \dfrac{2}{3} = \dfrac{7 \times \cancel{2}}{(4 \times \cancel{2}) \times 3} = \dfrac{7}{4 \times 3} = \dfrac{7}{12}$

Eliminate common factors before you multiply. Your work will look like this when you learn to do these operations mentally:

$$\frac{7}{\cancel{4}\cancel{8}} \times \frac{\cancel{2}^{1}}{3} = \frac{7}{12}$$

(b) $\dfrac{8}{12} \times \dfrac{3}{16} = \dfrac{\cancel{8} \times \cancel{3}}{(4 \times \cancel{3}) \times (\cancel{8} \times 2)} = \dfrac{1}{4 \times 2} = \dfrac{1}{8}$ or $\dfrac{\cancel{8}}{\cancel{12}_4} \times \dfrac{\cancel{3}}{\cancel{16}_2} = \dfrac{1}{8}$

(c) $\dfrac{\cancel{8}}{\cancel{32}_8} \times \dfrac{\cancel{4}^{1}}{\cancel{15}_5} = \dfrac{1}{40}$

(d) $\dfrac{\cancel{15}^{3}}{4} \times \dfrac{9}{\cancel{10}_2} = \dfrac{27}{8} = 3\dfrac{3}{8}$

(e) $\dfrac{\cancel{8}^{1}}{\cancel{2}_{1}} \times \dfrac{\cancel{2}^{1}}{\cancel{8}_{1}} = 1$

(f) $1\dfrac{1}{2} \times \dfrac{2}{5} = \dfrac{3}{\cancel{2}_{1}} \times \dfrac{\cancel{2}^{1}}{5} = \dfrac{3}{5}$

(g) $4 \times \dfrac{7}{8} = \dfrac{\cancel{4}^{1}}{1} \times \dfrac{7}{\cancel{8}_2} = \dfrac{7}{2} = 3\dfrac{1}{2}$

(h) $3\dfrac{5}{6} \times \dfrac{3}{10} = \dfrac{23}{\cancel{6}_2} \times \dfrac{\cancel{3}^{1}}{10} = \dfrac{23}{20} = 1\dfrac{3}{20}$

(i) $\dfrac{\cancel{4}^{1}}{\cancel{8}_{1}} \times \dfrac{\cancel{8}^{1}}{\cancel{4}_{1}} = 1$

Can you extend your new skills to solve these problems?

(a) $1\dfrac{4}{5} \times \dfrac{2}{3} \times \dfrac{1}{4} = $ _____

(b) $\left(\dfrac{1}{2}\right)^{2} = $ _____

(c) $\left(1\dfrac{2}{3}\right)^{3} = $ _____

(d) $\sqrt{\dfrac{16}{81}} = $ _____

Hint: $\sqrt{\dfrac{a}{b}} = \dfrac{\sqrt{a}}{\sqrt{b}}$

(continued)

Try them. An explanation is waiting in **25.**

24

Problem Set 2: Multiplication of Fractions

A. Multiply and reduce the answer to lowest terms:

$\dfrac{1}{2} \times \dfrac{1}{4} =$ _____ 　　　$\dfrac{2}{3} \times \dfrac{1}{6} =$ _____ 　　　$\dfrac{2}{5} \times \dfrac{2}{3} =$ _____

$\dfrac{3}{8} \times \dfrac{1}{3} =$ _____ 　　　$\dfrac{4}{5} \times \dfrac{1}{6} =$ _____ 　　　$\dfrac{5}{3} \times \dfrac{1}{2} =$ _____

$6 \times \dfrac{1}{2} =$ _____ 　　　$\dfrac{5}{6} \times \dfrac{3}{5} =$ _____ 　　　$\dfrac{8}{9} \times 3 =$ _____

$\dfrac{5}{16} \times \dfrac{8}{3} =$ _____ 　　　$\dfrac{11}{12} \times \dfrac{4}{15} =$ _____ 　　　$\dfrac{3}{7} \times \dfrac{3}{8} =$ _____

$\dfrac{8}{3} \times \dfrac{5}{12} =$ _____ 　　　$14 \times \dfrac{3}{4} =$ _____ 　　　$\dfrac{7}{8} \times \dfrac{13}{14} =$ _____

$\dfrac{5}{9} \times \dfrac{36}{25} =$ _____ 　　　$\dfrac{12}{8} \times \dfrac{15}{9} =$ _____ 　　　$\dfrac{32}{5} \times \dfrac{15}{16} =$ _____

$\dfrac{4}{7} \times \dfrac{49}{2} =$ _____ 　　　$\dfrac{16}{6} \times \dfrac{15}{28} =$ _____ 　　　$\dfrac{18}{5} \times \dfrac{10}{27} =$ _____

B. Multiply and reduce the answer to lowest terms:

$4\dfrac{1}{2} \times \dfrac{2}{3} =$ _____ 　　　　　$3\dfrac{1}{5} \times 1\dfrac{1}{4} =$ _____

$6 \times 1\dfrac{1}{3} =$ _____ 　　　　　$\dfrac{3}{8} \times 3\dfrac{1}{2} =$ _____

$2\dfrac{1}{6} \times 1\dfrac{1}{2} =$ _____ 　　　　　$7\dfrac{3}{4} \times 8 =$ _____

$\dfrac{5}{7} \times 1\dfrac{7}{15} =$ _____ 　　　　　$1\dfrac{2}{9} \times \dfrac{3}{11} =$ _____

$4\dfrac{3}{5} \times 15 =$ _____ 　　　　　$3\dfrac{3}{8} \times 1\dfrac{7}{9} =$ _____

$10\dfrac{5}{6} \times 3\dfrac{3}{10} =$ _____ 　　　　　$4\dfrac{5}{11} \times \dfrac{2}{7} =$ _____

$34 \times 2\dfrac{3}{17} =$ _____ 　　　　　$9\dfrac{7}{8} \times \dfrac{4}{5} =$ _____

$7\dfrac{9}{10} \times 1\dfrac{1}{4} =$ _____ 　　　　　$14 \times 3\dfrac{1}{3} =$ _____

$11\frac{6}{7} \times \frac{7}{8} =$ _____ $5\frac{1}{6} \times 2\frac{3}{5} =$ _____

$18 \times 1\frac{5}{27} =$ _____ $3\frac{1}{5} \times 1\frac{7}{8} =$ _____

C. Solve:

$(\frac{2}{3})^2 =$ _____ $(\frac{1}{2})^4 =$ _____ $(\frac{3}{5})^3 =$ _____

$(3\frac{1}{5})^2 =$ _____ $(4\frac{1}{2})^3 =$ _____ $\sqrt{\frac{9}{16}} =$ _____

$\sqrt{\frac{4}{49}} =$ _____ $\sqrt{\frac{16}{25}} =$ _____ $\sqrt{\frac{25}{64}} =$ _____

$\sqrt{\frac{81}{121}} =$ _____ $\frac{1}{4} \times \frac{2}{3} \times \frac{2}{5} =$ _____

$\frac{5}{12} \times \frac{3}{4} \times \frac{8}{15} =$ _____ $2\frac{1}{2} \times \frac{3}{5} \times \frac{8}{9} =$ _____

$5\frac{1}{3} \times 2\frac{1}{4} \times 1\frac{2}{3} =$ _____ $12\frac{2}{3} \times 5\frac{1}{4} \times \frac{9}{19} =$ _____

D. Brain Boosters

1. A jet plane cruises at 450 mph. How far does it travel in $3\frac{2}{5}$ hours?

2. John ate $\frac{1}{3}$ of a whole apple pie. Later Bert ate $\frac{3}{4}$ of the remainder. What part of the total pie did Bert eat?

3. The scale on a map is 1 centimeter equals $12\frac{1}{2}$ kilometers. What actual distance is represented by a map distance of $8\frac{4}{5}$ centimeters?

4. A boy spends $\frac{1}{3}$ of his savings and loses $\frac{2}{3}$ of the remainder. He then has 12¢. How much money did he start with?

5. Show that this problem is correctly done:

$$\frac{18534}{9267} \times \frac{17469}{5823} = \frac{34182}{5697}$$

Notice that each fraction contains all nine nonzero digits.

6. What is the area in square miles of a farm $1\frac{5}{16}$ miles long by $\frac{2}{3}$ mile wide?

7. What is the total amount of medication in 6 pills each containing $2\frac{3}{4}$ milligrams?

The answers to these problems are on page 227. When you have had the practice you need, either return to the preview test on page 76 or continue in **26** with the study of the division of fractions.

25

(a) $1\frac{4}{5} \times \frac{2}{3} \times \frac{1}{4} = \frac{9}{5} \times \frac{2}{3} \times \frac{1}{4} = \frac{\cancel{9}^3 \times \cancel{2} \times 1}{5 \times \cancel{3} \times \cancel{4}_2} = \frac{3}{10}$

Multiplication of three or more fractions involves nothing new. Be sure to change all mixed numbers to improper fractions before multiplying.

(b) $\left(\frac{1}{2}\right)^2 = \frac{1}{2} \times \frac{1}{2} = \frac{1 \times 1}{2 \times 2} = \frac{1}{4}$ (Easy, right?)

(c) $\left(1\frac{2}{3}\right)^3 = \left(\frac{5}{3}\right)^3 = \frac{5}{3} \times \frac{5}{3} \times \frac{5}{3} = \frac{5 \times 5 \times 5}{3 \times 3 \times 3} = \frac{125}{27}$

Again, you must change any mixed numbers to improper fractions <u>before</u> you multiply.

(d) $\sqrt{\frac{16}{81}} = \frac{\sqrt{16}}{\sqrt{81}} = \frac{4}{9}$ Check: $\frac{4}{9} \times \frac{4}{9} = \frac{4 \times 4}{9 \times 9} = \frac{16}{81}$

Square roots are not difficult if both numerator and denominator are perfect squares. Otherwise you must use the Table of Square Roots and just divide decimals as shown in Chapter 3.

Now turn to **24** for a set of practice problems on multiplication.

Division

26

Addition and multiplication are both reversible arithmetic operations. For example,

$$2 \times 3 \text{ and } 3 \times 2 \text{ both equal } 6$$
$$4 + 5 \text{ and } 5 + 4 \text{ both equal } 9$$

The order in which you add addends or multiply factors is not important. This fact is called the <u>commutative property of addition and multiplication</u>.

In subtraction and division this kind of exchange is <u>not</u> allowed and because of this many people find these operations very troublesome.

7 − 5 = 2 but 5 − 7 is <u>not</u> equal to 2 and is not even a counting number.

8   4 = 2 but 4 ÷ 8 is <u>not</u> equal to 2 and is not a whole number.

In the division of fractions it is particularly important that you set up the process correctly.

Are these four numbers equal?

$$\text{"8 divided by 4"} \qquad 8\overline{)4} \qquad 8 \div 4 \qquad \frac{4}{8}$$

Choose an answer: (a)   Yes   Go to **28.**

                      (b)   No   Go to **29.**

27

The divisor is $\frac{1}{2}$. The division $5 \div \frac{1}{2}$ is read "5 divided by $\frac{1}{2}$" and it asks how many $\frac{1}{2}$ unit lengths are included in a length of 5 units.

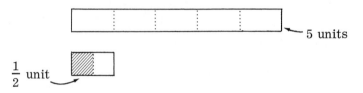

$\frac{1}{2}$ unit

Division is defined in terms of multiplication.

$$8 \div 4 = \square$$

asks that you find a number $\square$ such that $8 = \square \times 4$. It is easy to see that $\square = 2$.

Division is defined in exactly the same way with fractions.

$$5 \div \frac{1}{2} = \square$$

asks that you find a number $\square$ such that $5 = \square \times \frac{1}{2}$.

Working backward from this multiplication and using the diagram above, find the answer to $5 \div \frac{1}{2}$. Hop ahead to **30** to continue.

28

Your answer is incorrect. Be very careful about this.

$$8 \div 4 \text{ is read "8 divided by 4" and written } 4\overline{)8} \text{ or } \frac{8}{4}$$

4 is the divisor

In the problem $8 \div 4$ you are being asked to divide a set of 8 objects into sets of 4 objects. The divisor (4) is the denominator or bottom number of the fraction.

In the division $5 \div \dfrac{1}{2}$ which number is the divisor? Check your answer in **27.**

29

Right you are. "8 divided by 4" is written $8 \div 4$. We can also write this as $4\overline{)8}$ or $\dfrac{8}{4}$. In all of these the divisor is 4 and you are being asked to divide a set of 8 objects into sets of 8 objects.

In the division $5 \div \dfrac{1}{2}$ which number is the divisor? Check your answer in **27.**

30

$$5 \div \frac{1}{2} = 10$$

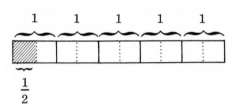

There are ten $\dfrac{1}{2}$ unit lengths contained in the 5 unit length.

Using a drawing of this sort to solve a division problem is difficult and clumsy. We need a simple rule. Here it is:

> To divide by a fraction, invert the divisor and multiply.

Example 1

$$\overset{\text{multiply}}{5 \div \frac{1}{2} = 5 \times \frac{2}{1}} = 5 \times 2 = 10$$

The divisor $\dfrac{1}{2}$ has been inverted.

To invert a fraction simply means to switch top and bottom. Inverting $\dfrac{2}{3}$ gives $\dfrac{3}{2}$. Inverting $\dfrac{1}{5}$ gives $\dfrac{5}{1}$. Inverting 7 gives $\dfrac{1}{7}$.

Example 2

$$\frac{3}{5} \div \frac{2}{3} = \frac{3}{5} \times \frac{3}{2} = \frac{3 \times 3}{5 \times 2} = \frac{9}{10}$$

We have converted a division problem that is difficult to picture into a simple multiplication. The final, and very important, step in every division is checking the answer.

$$\text{If } \frac{3}{5} \div \frac{2}{3} = \frac{9}{10}, \text{ then } \frac{3}{5} = \frac{2}{3} \times \frac{9}{10}.$$

$$\text{Check: } \frac{\cancel{2}^{1}}{\cancel{3}} \times \frac{\cancel{9}^{3}}{\cancel{10}5} = \frac{3}{5}$$

Why does this "invert and multiply" process work? The division $\frac{3}{5} \div \frac{2}{3} = \square$ means that there is some number $\square$ such that $\frac{3}{5} = \square \times \frac{2}{3}$. Multiply both sides of the last equation by $\frac{3}{2}$:

$$(\frac{3}{5}) \times \frac{3}{2} = (\square \times \frac{2}{3}) \times \frac{3}{2}$$

But we can do multiplication in any order we wish and we know that $\frac{2}{3} \times \frac{3}{2} = \frac{6}{6} = 1$, so $\frac{3}{5} \times \frac{3}{2} = \square$. Our answer, the unknown number we have labeled $\square$, is simply the product of the dividend $\frac{3}{5}$ and the inverted divisor $\frac{3}{2}$.

Try this one:

$$\frac{7}{8} \div \frac{3}{2} = \underline{\hspace{2cm}}$$

Solve it by inverting the divisor and multiplying. Check your answer in **31.**

Ours is not to reason why, only to invert and multiply. Is that it?

Wise guy. Invert and multiply is the rule for dividing fractions, and it will always work, but it's a mistake to do arithmetic without reasoning why it works. See the box on the next page for an explanation.

31

$$\frac{7}{8} \div \frac{3}{2} = \frac{7}{8} \times \frac{2}{3} = \frac{7 \times \cancel{2}^{1}}{\cancel{4}\cancel{8} \times 3} = \frac{7}{12} \qquad \text{Check:} \quad \frac{\cancel{3}^{1}}{2} \times \frac{7}{\cancel{12}_{4}} = \frac{7}{8}$$

The chief source of confusion in dividing fractions is deciding which fraction to invert. It will help if you:

●1. First put every division problem in the form

(dividend) ÷ (divisor)

Then invert the divisor and multiply to obtain the quotient.

●2. Check your answer by multiplying. The product

(divisor) × (quotient or answer)

should equal the dividend.

Here are a few problems to test your understanding.

(a) $\dfrac{2}{5} \div \dfrac{3}{8} =$ _____

(b) $\dfrac{7}{40} \div \dfrac{21}{25} =$ _____

(c) $3\dfrac{3}{4} \div \dfrac{5}{2} =$ _____

(d) $4\dfrac{1}{5} \div 1\dfrac{4}{10} =$ _____

(e) $3\dfrac{2}{3} \div 3 =$ _____

(f) Divide $\dfrac{3}{4}$ by $2\dfrac{5}{8}$. _____

(g) Divide 8 by $\dfrac{1}{2}$. _____

(h) Divide $1\dfrac{1}{4}$ by $1\dfrac{7}{8}$. _____

Work carefully, check each answer, then turn to **32** for our worked solutions.

WHY DO WE INVERT AND MULTIPLY TO DIVIDE FRACTIONS?

The division 8 ÷ 4 can be written $\dfrac{8}{4}$. Similarly, $\dfrac{1}{2} \div \dfrac{2}{3}$ can be written

$$\frac{\dfrac{1}{2}}{\dfrac{2}{3}}$$

To simplify this fraction multiply by $\dfrac{\dfrac{3}{2}}{\dfrac{3}{2}}$.

$$\frac{\frac{1}{2}}{\frac{2}{3}} = \frac{\frac{1}{2} \times \frac{3}{2}}{\frac{2}{3} \times \frac{3}{2}} = \frac{\frac{1}{2} \times \frac{3}{2}}{1} = \frac{1}{2} \times \frac{3}{2}$$

Notice the second fraction here ($\frac{3}{2}$) is the original denominator inverted. Also, in the denominator $\frac{2}{3} \times \frac{3}{2} = \frac{2 \times 3}{3 \times 2} = \frac{6}{6} = 1$.
Therefore, $\frac{1}{2} \div \frac{2}{3} = \frac{1}{2} \times \frac{3}{2}$. We have inverted the fraction $\frac{2}{3}$ and multiplied by it.

32

(a) $\frac{2}{5} \div \frac{3}{8} = \frac{2}{5} \times \frac{8}{3} = \frac{16}{15} = 1\frac{1}{15}$ Check: $\frac{\cancel{8}^1}{\cancel{8}_1} \times \frac{\cancel{16}^2}{\cancel{15}_5} = \frac{2}{5}$

Remember $\frac{3}{8} \times \frac{16}{15} = \frac{3 \times 16}{8 \times 15} = \frac{\cancel{3} \times 2 \times \cancel{8}}{\cancel{8} \times \cancel{3} \times 5} = \frac{2}{5}$

(b) $\frac{7}{40} \div \frac{21}{25} = \frac{7}{40} \times \frac{25}{21} = \frac{5}{24}$ Check: $\frac{\cancel{21}^7}{\cancel{25}_5} \times \frac{\cancel{5}^1}{\cancel{24}_8} = \frac{7}{40}$

(c) $3\frac{3}{4} \div \frac{5}{2} = \frac{15}{4} \div \frac{5}{2} = \frac{\cancel{15}^3}{\cancel{4}_2} \times \frac{\cancel{2}^1}{\cancel{5}_1} = \frac{3}{2} = 1\frac{1}{2}$ Check: $\frac{5}{2} \times \frac{3}{2} = \frac{15}{4} = 3\frac{3}{4}$

(d) $4\frac{1}{5} \div 1\frac{4}{10} = \frac{21}{5} \div \frac{14}{10} = \frac{\cancel{21}^3}{\cancel{5}_1} \times \frac{\cancel{10}^{\cancel{2}1}}{\cancel{14}\cancel{x}_1} = 3$

Check: $1\frac{4}{10} \times 3 = \frac{14}{10} \times \frac{3}{1} = \frac{42}{10} = 4\frac{2}{10} = 4\frac{1}{5}$

(e) $3\frac{2}{3} \div 3 = \frac{11}{3} \div \frac{3}{1} = \frac{11}{3} \times \frac{1}{3} = \frac{11}{9} = 1\frac{2}{9}$

Check: $3 \times \frac{11}{9} = \frac{3}{1} \times \frac{11}{9} = \frac{11}{3} = 3\frac{2}{3}$

(f) $\frac{3}{4} \div 2\frac{5}{8} = \frac{3}{4} \div \frac{21}{8} = \frac{\cancel{3}^1}{\cancel{4}1} \times \frac{\cancel{8}^2}{\cancel{21}_7} = \frac{2}{7}$ Check: $2\frac{5}{8} \times \frac{2}{7} = \frac{21}{8} \times \frac{2}{7} = \frac{3}{4}$

(g) $8 \div \frac{1}{2} = \frac{8}{1} \times \frac{2}{1} = 16$ Check: $\frac{1}{2} \times 16 = 8$

(h) $1\frac{1}{4} \div 1\frac{7}{8} = \frac{5}{4} \div \frac{15}{8} = \frac{\cancel{5}^1}{\cancel{4}_1} \times \frac{\cancel{8}^2}{\cancel{15}_3} = \frac{2}{3}$

Check: $1\dfrac{7}{8} \times \dfrac{2}{3} = \dfrac{15}{8} \times \dfrac{2}{3} = \dfrac{5}{4} = 1\dfrac{1}{4}$

Turn to **33** for a set of practice problems on dividing fractions.

RATIO AND PROPORTION

A <u>ratio</u> is a comparison of the sizes of two quantities of the same kind. It is a single number, usually written as a fraction. For example, the steepness of a hill can be written as the ratio of its height to its horizontal extent.

$$\text{steepness} = \frac{10 \text{ ft}}{80 \text{ ft}} = \frac{1}{8}$$

10 ft

80 ft

The ratio of the circumference of any circle to its diameter is approximately 3.14. This ratio is used so often in mathematics that it has been given a special symbol, the Greek letter π (pronounced "pie").

A <u>proportion</u> is a statement that two ratios are equal. It is an equation or a statement in words and is most interesting when one of the ratios is incomplete. For example, if the ratio of Joe's height to John's height is 7 to 8 and John is 6 feet tall, how tall is Joe?

$$\text{ratio of heights} = \frac{\text{Joe's height}}{\text{John's height}} = \frac{7}{8} = \frac{\boxed{?}}{6 \text{ ft}}$$

This equation is a proportion. It is always true that if two fractions are equal

$$\frac{a}{b} = \frac{c}{d} \quad \text{then} \quad \frac{a \times d}{b \times d} = \frac{c \times b}{d \times b}$$

Or, since the denominators are equal, $a \times d = c \times b$. In any proportion the cross products are equal.

$$\frac{a}{b} = \frac{c}{d} \qquad a \times d = c \times b$$

Therefore, $\boxed{?} \times 8 = 7 \text{ ft} \times 6 \text{ ft}$

$$\boxed{?} = \frac{7 \text{ ft} \times 6 \text{ ft}}{8} = \frac{42}{8} \text{ ft} = 5.25 \text{ ft}$$

The following problem shows how useful proportions can be.

A telephone pole cases a shadow 9 ft long. A yard
stick beside it casts a shadow 16 inches long. What
is the height of the telephone pole?

$$\frac{\boxed{?}}{9 \text{ ft}} = \frac{36 \text{ in.}}{16 \text{ in.}}$$

$$\boxed{?} = \frac{36 \text{ in.} \times 9 \text{ ft}}{16 \text{ in.}} = \frac{81}{4} \text{ ft} = 20.25 \text{ ft}$$

Try these problems to exercise your understanding of ratio
and proportion.

1. In the college 2430 students are male and 2970 are female.
   How many women students would you expect to find in a class
   where there are 18 men?

2. In the first 12 games of the season, our star basketball player
   scored 252 points. How many can he expect to score in the
   entire 22 game season?

3. A car travels 132 miles on 8 gallons of gasoline. How far
   would you expect it to travel on a full tank of 12 gallons?

The answers to these problems are on page 228.

33

Problem Set 3: Dividing Fractions

A. Divide and reduce the answer to lowest terms:

$\dfrac{5}{6} \div \dfrac{1}{2} =$ _____ $\dfrac{3}{4} \div \dfrac{3}{7} =$ _____ $6 \div \dfrac{2}{3} =$ _____

$\dfrac{1}{2} \div 6 =$ _____ $\dfrac{5}{12} \div \dfrac{4}{3} =$ _____ $\dfrac{4}{18} \div \dfrac{1}{2} =$ _____

$8 \div \dfrac{1}{3} =$ _____ $\dfrac{7}{20} \div \dfrac{4}{5} =$ _____ $\dfrac{6}{13} \div \dfrac{3}{4} =$ _____

$3 \div \dfrac{2}{5} =$ _____ $\dfrac{1}{2} \div \dfrac{1}{2} =$ _____ $\dfrac{1}{2} \div \dfrac{1}{3} =$ _____

$\dfrac{3}{14} \div \dfrac{6}{5} =$ _____ $\dfrac{3}{5} \div \dfrac{1}{3} =$ _____ $\dfrac{3}{4} \div \dfrac{5}{16} =$ _____

B. Divide and reduce the answer to lowest terms:

$1\dfrac{1}{2} \div \dfrac{1}{6} =$ _____ $2\dfrac{3}{4} \div \dfrac{3}{8} =$ _____ $6 \div 1\dfrac{1}{2} =$ _____

$2\dfrac{1}{4} \div 3 =$ _____ $3\dfrac{1}{7} \div 2\dfrac{5}{14} =$ _____ $8\dfrac{2}{5} \div 1\dfrac{2}{5} =$ _____

$3\dfrac{1}{2} \div 2 =$ _____ $4\dfrac{1}{2} \div 1\dfrac{3}{4} =$ _____ $6\dfrac{2}{5} \div 5\dfrac{1}{3} =$ _____

$10 \div 1\dfrac{1}{5} =$ _____ $4\dfrac{1}{6} \div 3\dfrac{1}{3} =$ _____ $15\dfrac{5}{6} \div 9\dfrac{1}{2} =$ _____

$7\dfrac{1}{7} \div 8\dfrac{1}{3} =$ _____ $11\dfrac{2}{3} \div 2\dfrac{2}{9} =$ _____ $1\dfrac{1}{5} \div 1\dfrac{1}{2} =$ _____

C. Divide:

$\dfrac{8}{\dfrac{1}{2}} =$ $\dfrac{\dfrac{3}{4}}{2} =$ $\dfrac{\dfrac{2}{3}}{6} =$

$\dfrac{2\dfrac{1}{3}}{\dfrac{3}{4}} =$ $\dfrac{12}{\dfrac{2}{3}} =$ $\dfrac{15}{\dfrac{3}{5}} =$

Divide $\dfrac{3}{4}$ by $\dfrac{7}{8}$.   Divide 2 by $\dfrac{1}{3}$.

Divide $1\dfrac{7}{8}$ by $\dfrac{3}{2}$.   Divide $1\dfrac{1}{2}$ by $7\dfrac{1}{4}$.

Divide 5 by $\dfrac{2}{7}$.   Divide $11\dfrac{1}{3}$ by $\dfrac{2}{3}$.

D. Brain Boosters

1. Mr. Megabuck drove his new Cardiac sedan $34\dfrac{1}{2}$ miles and used $3\dfrac{1}{4}$ gallons of gasoline. How many miles per gallon did he average?

2. The product of two numbers is $14\dfrac{2}{5}$. One of the numbers is $3\dfrac{1}{5}$. What is the other number?

3. If you drive $59\dfrac{1}{2}$ miles in $1\dfrac{3}{4}$ hours, what is your average speed?

4. A length of cloth $6\frac{7}{8}$ yards long is to be divided into 5 equal pieces. What will be the length of each piece?

The answers to these problems are on page 228. When you have had the practice you need either return to the preview test on page 76 or continue in **34** with the addition and subtraction of fractions.

Addition and Subtraction

34

At heart, adding fractions is a matter of counting:

$$\frac{1}{5} + \frac{3}{5} = \frac{1+3}{5} = \frac{4}{5}$$

$\frac{1}{5}$ ◁— 1 fifth

\+

$\frac{3}{5}$ ◁— + 3 fifths

=

$\frac{4}{5}$ ◁— 4 fifths (count them)

You try this one:

$$\frac{2}{7} + \frac{3}{7} = \underline{\hspace{3cm}}$$

Check your answer in **35.**

35

$$\frac{2}{7} + \frac{3}{7} = \frac{2+3}{7} = \frac{5}{7}$$

$\frac{2}{7}$ ◁— 2 sevenths

\+

$\frac{3}{7}$ ◁— + 3 sevenths

=

$\frac{5}{7}$ ◁— 5 sevenths or $\frac{5}{7}$

Fractions having the same denominator are called <u>like fractions</u>. In the problem above, $\frac{2}{7}$ and $\frac{3}{7}$ both have the denominator 7 and are like fractions. Adding like fractions is easy.

To add like fractions:

- First, add the numerators to find the numerator of the sum.

- Second, use the denominator the fractions have in common as the denominator of the sum.

$$\frac{2}{9} + \frac{5}{9} = \frac{2+5}{9} = \frac{7}{9} \quad \text{(add numerators)} \atop \text{(same denominator)}$$

Adding three or more fractions presents no special problems.

$$\frac{3}{12} + \frac{1}{12} + \frac{5}{12} = \underline{\hspace{3cm}}$$

Add the fractions as shown above, then turn to **36**.

36

$$\frac{3}{12} + \frac{1}{12} + \frac{5}{12} = \frac{3+1+5}{12} = \frac{9}{12} = \frac{3}{4}$$

Notice that we reduce the sum to lowest terms.
Try these problems for practice:

(a) $\dfrac{1}{8} + \dfrac{3}{8} = \underline{\hspace{2cm}}$

(b) $\dfrac{7}{9} + \dfrac{5}{9} = \underline{\hspace{2cm}}$

(c) $2\dfrac{1}{2} + 3\dfrac{3}{2} = \underline{\hspace{2cm}}$

(d) $2 + 3\dfrac{1}{2} = \underline{\hspace{2cm}}$

(e) $\dfrac{1}{7} + \dfrac{4}{7} + \dfrac{5}{7} + 1\dfrac{2}{7} + \dfrac{8}{7} = \underline{\hspace{2cm}}$

(f) $\dfrac{3}{5} + 1\dfrac{1}{5} + 3 = \underline{\hspace{2cm}}$

Go to **38** to check your work.

37

$$\frac{3}{4} + \frac{2}{3} = \frac{9}{12} + \frac{8}{12} = \frac{17}{12} = 1\frac{5}{12}$$

We change the original fractions to equivalent fractions with the same denominator and then add as before.

How do you know what number to use as the new denominator? In general, you cannot simply guess at the best new denominator. We need a method for finding it from the denominators of the fractions to be added. The new denominator we want is called the Least Common Multiple or LCM.

Every whole number has a set of <u>multiples</u> associated with it. We can find the multiples of any given number by multiplying it by each of the whole numbers in turn. For example,

> The multiples of 2 are 2, 4, 6, 8, 10, 12, . . .
> (Multiply each integer 1, 2, 3, 4, . . . by 2.)
>
> The multiples of 3 are 3, 6, 9, 12, 15, 18, . . .
> (Multiply each integer 1, 2, 3, 4, . . . by 3.)
>
> The multiples of 4 are 4, 8, 12, 16, 20, 24, . . .

Find the first few multiples of the following:

(a) 5   (b) 7   (c) 12   (d) 8

Our answers are in **39.**

38

(a) $\dfrac{1}{8} + \dfrac{3}{8} = \dfrac{1+3}{8} = \dfrac{4}{8} = \dfrac{1}{2}$ (reduced to lowest terms)

(b) $\dfrac{7}{9} + \dfrac{5}{9} = \dfrac{7+5}{9} = \dfrac{12}{9} = \dfrac{4}{3} = 1\dfrac{1}{3}$

(c) $2\dfrac{1}{2} + 3\dfrac{3}{2} = 2 + 3 + \dfrac{1}{2} + \dfrac{3}{2} = 2 + 3 + \dfrac{4}{2} = 2 + 3 + 2 = 7$

(Add the whole number parts first.)

(d) $2 + 3\dfrac{1}{2} = 2 + 3 + \dfrac{1}{2} = 5\dfrac{1}{2}$ (Remember, $3\dfrac{1}{2}$ means $3 + \dfrac{1}{2}$.)

(e) $\dfrac{1}{7} + \dfrac{4}{7} + \dfrac{5}{7} + 1\dfrac{2}{7} + \dfrac{8}{7} = \dfrac{1}{7} + \dfrac{4}{7} + \dfrac{5}{7} + \dfrac{9}{7} + \dfrac{8}{7} = \dfrac{1+4+5+9+8}{7} =$

$\dfrac{27}{7} = 3\dfrac{6}{7}$

(f) $\dfrac{3}{5} + 1\dfrac{1}{5} + 3 = 1 + 3 + \dfrac{3}{5} + \dfrac{1}{5} = 4 + \dfrac{4}{5} = 4\dfrac{4}{5}$

(Add the whole number parts first.)

How do we add fractions whose denominators are not the same?

$$\dfrac{2}{3} + \dfrac{3}{4} = \underline{\qquad\qquad} ?$$

The problem is to find a simple numeral that names this new number. One way to find it is to change these fractions to equivalent fractions with the same denominator. (Equivalent fractions are discussed on page 85 in frame **10.**)

$$\dfrac{3}{4} = \dfrac{3 \times 3}{4 \times 3} = \dfrac{9}{12} \qquad\qquad \dfrac{2}{3} = \dfrac{2 \times 4}{3 \times 4} = \dfrac{8}{12}$$

Now that $\dfrac{2}{3}$ and $\dfrac{3}{4}$ have been changed to equivalent fractions you can add them. Do it now and then go to **37.**

39

(a) The multiples of 5 are 5, 10, 15, 20, 25, . . .
(b) The multiples of 7 are 7, 14, 21, 28, 35, . . .
(c) The multiples of 12 are 12, 24, 36, 48, 60, 72, . . .
(d) The multiples of 8 are 8, 16, 24, 32, 40, 48, 56, 64, 72, . . .

Notice that 24, 48, and 72 are multiples of both 8 and 12. They are the first three <u>common</u> <u>multiples</u> of 8 and 12.
List a few of the multiples common to 3 and 5. Go to **40** to check your answer.

40

The multiples of 3 are 3, 6, 9, 12, 15, 18, 21, 24, 27, 30, . . .

The multiples of 5 are 5, 10, 15, 20, 25, 30, 35, . . .

The multiples 3 and 5 have in common are 15, 30, 45, 60, and so on.

List several of the common multiples of the following pairs of numbers:

(a) 4 and 6   (b) 12 and 16   (c) 8 and 10

For each pair make two lists of multiples and find the numbers which appear on both lists. Look in **41** for the correct answers.

41

(a) The multiples of 4 and 6 are:

      4: 4, 8, 12, 16, 20, 24, 28, 32, 36, 40, 44, 48 . . .

      6: 6, 12, 18, 24, 30, 36, 42, 48, 54, . . .

The common multiples are 12, 24, 36, 48, and so on.

(b) The common multiples of 12 and 16 are 48, 96, 144, 192, and so on.

(c) The common multiples of 8 and 10 are 40, 80, 120, 160, and so on.

The <u>Least</u> <u>Common</u> <u>Multiple</u> or LCM is the smallest common multiple in the set of common multiples for a pair of numbers. For example, the Least Common Multiple of 4 and 6 in (a) above is 12. The LCM is the smallest number that both 4 and 6 will divide evenly.

      The LCM of 12 and 16 is 48.
      The LCM of 8 and 10 is 40.

    What is the LCM of 8 and 12? _____

The answer is in **42.**

42

    The LCM or Least Common Multiple of 8 and 12 is 24. It is the smallest number that both 8 and 12 will divide with zero remainder.

    Making lists of multipliers is the most direct way of finding the LCM, but it is slow and time-consuming. Here is a short-cut method we call the <u>LCM</u> <u>Finder</u>.

Example

■1. Write each denominator as a product of primes.

To add $\dfrac{7}{60}$ + $\dfrac{11}{72}$ find the LCM of 60 and 72:

$$60 = 2^2 \times 3^1 \times 5^1$$
$$72 = 2^3 \times 3^2$$

■2. Write each base prime that appears in either number.

    2   3   5

■3. Attach to each prime the largest exponent that appears on it in either number.

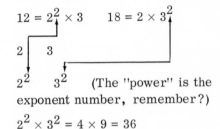

2^3 3^2 5^1

from   from   from
2^3 in 72 3^2 in 72 5^1 in 60

■4. Multiply to find the LCM. Use this product as the new denominator.

$$LCM = 2^3 \times 3^2 \times 5^1$$
$$= 8 \times 9 \times 5$$
$$= 360$$

Try it on this pair of numbers. Find the LCM of 12 and 18. Check your work in **43**.

43

Factor 12 and 18:

$$12 = 2^2 \times 3 \qquad 18 = 2 \times 3^2$$

Write the primes that appear as factors of 12 and 18:

2 | 3

Attach the highest powers of each that appear in either number:

2^2 3^2 (The "power" is the exponent number, remember?)

Multiply:

$$2^2 \times 3^2 = 4 \times 9 = 36$$

LCM = 36

A bit of practice will groove the idea into place. Find the Least Common Multiple of each of the following pairs of numbers.

(a) 15 and 6   (b) 20 and 24   (c) 54 and 180

(d) 525 and 90   (e) 8 and 18   (f) 140 and 50

Answers in **44**.

44

(a) $15 = 3^1 \times 5^1$ $6 = 2^1 \times 3^1$
    $LCM = 2^1 \times 3^1 \times 5^1 = 30$

(b) $20 = 2^2 \times 5^1$ $24 = 2^3 \times 3^1$
    $LCM = 2^3 \times 3^1 \times 5^1 = 8 \times 3 \times 5 = 120$

(c) $54 = 2^1 \times 3^3$ $180 = 2^2 \times 3^2 \times 5^1$
    $LCM = 2^2 \times 3^3 \times 5^1 = 4 \times 27 \times 5 = 540$

(d) $525 = 3^1 \times 5^2 \times 7$ $90 = 2^1 \times 3^2 \times 5^1$
    $LCM = 2^1 \times 3^2 \times 5^2 \times 7 = 2 \times 9 \times 25 \times 7 = 3150$

(e) $8 = 2^3$ $18 = 2^1 \times 3^2$
    $LCM = 2^3 \times 3^2 = 8 \times 9 = 72$

(f) $140 = 2^2 \times 5^1 \times 7^1$ $50 = 2^1 \times 5^2$
      LCM $= 2^2 \times 5^2 \times 7^1 = 4 \times 25 \times 7 = 700$

The process is exactly the same with three or more numbers. For example, the LCM of 24, 90, and 75 is found this way:

$24 = 2^3 \times 3^1$ $90 = 2^1 \times 3^2 \times 5^1$ $75 = 3^1 \times 5^2$

LCM $= 2^3 \times 3^2 \times 5^2 = 8 \times 9 \times 25 = 1800$

Find the sum of $\dfrac{3}{18} + \dfrac{5}{12}$. Write both numbers as equivalent fractions with the LCM of 18 and 12 as the denominator, then add. Our solution is in **45**.

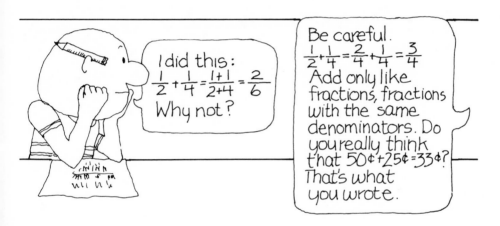

45

$18 = 2^1 \times 3^2$ $12 = 2^2 \times 3$
LCM $= 2^2 \times 3^2 = 4 \times 9 = 36$

$\dfrac{3}{18} = \dfrac{?}{36} = \dfrac{3 \times 2}{18 \times 2} = \dfrac{6}{36}$ $\dfrac{5}{12} = \dfrac{?}{36} = \dfrac{5 \times 3}{12 \times 3} = \dfrac{15}{36}$

Add: $\dfrac{3}{18} + \dfrac{5}{12} = \dfrac{6}{36} + \dfrac{15}{36} = \dfrac{6 + 15}{36} = \dfrac{21}{36} = \dfrac{7}{12}$ (in lowest terms)

To add fractions:

■1. Add the whole number parts of the fractions.

■2. Find the LCM of the denominators of the fractions to be added.

■3. Write the fractions so that they are equivalent fractions with the LCM as denominator.

■4. The numerator of the sum is the sum of the numerators and the denominator of the sum is the LCM.

■5. Reduce to lowest terms.

This way of adding fractions may seem rather long and involved. It is involved, but it is the only sure way to arrive at the answer. It is new to you and you will need lots of practice at it before it comes quickly. Take each problem step by step, work slowly at first, and gradually you will become very skilled at adding fractions.

Practice by adding the following:

(a) $\dfrac{5}{18} + \dfrac{5}{16}$

(b) $\dfrac{7}{12} + \dfrac{3}{16}$

(c) $\dfrac{10}{32} + \dfrac{4}{30}$

(d) $\dfrac{3}{7} + \dfrac{4}{5}$

(e) $1\dfrac{2}{15} + \dfrac{5}{9}$

(f) $\dfrac{1}{2} + \dfrac{5}{6} + \dfrac{3}{4}$

(g) $2\dfrac{5}{12} + 1\dfrac{1}{9} + 2\dfrac{3}{8}$

(h) $\dfrac{13}{16} + 2\dfrac{1}{18} + 1\dfrac{11}{12}$

The worked solutions are in **48**.

46

$$\frac{3}{8} - \frac{1}{8} = \frac{3-1}{8} = \frac{2}{8} = \frac{1}{4}$$

Easy enough? If the denominators are the same we subtract numerators and write this difference over the common denominator. Try another:

$$\frac{3}{4} - \frac{1}{5} = \underline{\qquad}$$

Our solution is in **47**.

47

The LCM of 4 and 5 is 20.

$$\frac{3}{4} - \frac{?}{20} = \frac{3 \times 5}{4 \times 5} = \frac{15}{20} \qquad \frac{1}{5} - \frac{?}{20} = \frac{1 \times 4}{5 \times 4} = \frac{4}{20}$$

$$\text{Then } \frac{3}{4} - \frac{1}{5} = \frac{15}{20} - \frac{4}{20} = \frac{11}{20}$$

When the two fractions have different denominators we must change them to equivalent fractions with the LCM as denominator before subtracting.

If you have not yet learned how to find the Least Common Multiple (LCM) turn to **41.** Otherwise, continue by finding the following differences.

(a) $3\dfrac{1}{4} - 1\dfrac{1}{12} =$ _____

(b) $10 - 3\dfrac{5}{16} =$ _____

(c) $4\dfrac{7}{18} - 2\dfrac{11}{12} =$ _____

(d) $2\dfrac{6}{32} - 1\dfrac{1}{6} =$ _____

Check your answers in **49.**

WHAT IS THE LCM?

An LCM is not a Little Crooked Martian, but a Least Common Multiple.

A multiple of a whole number is a number evenly divisible by it.

4, 8, 12, 16, 20, 24, . . . are all multiples of 4

A pair of numbers will have common multiples.

The multiples of 4 are 4, 8, 12, 16, 20, 24, . . .

The multiples of 8 are 8, 16, 24, 32, . . .

The common multiples of 4 and 8 are 8, 16, 24, and so on.

The smallest common multiple or least common multiple or LCM is the smallest of all the common multiples of a pair of whole numbers.

The LCM of 4 and 8 is 8. The LCM 8 is the smallest whole number that is evenly divisible by both 4 and 8.

48

(a) $18 = 2^1 \times 3^2 \qquad 16 = 2^4$

The LCM of 18 and 16 is $2^4 \times 3^2 = 144.$

$$\frac{5}{18} = \frac{?}{144} = \frac{5 \times 8}{18 \times 8} = \frac{40}{144} \qquad \frac{5}{16} = \frac{?}{144} = \frac{5 \times 9}{16 \times 9} = \frac{45}{144}$$

$$\frac{5}{18} + \frac{5}{16} = \frac{40}{144} + \frac{45}{144} = \frac{40 + 45}{144} = \frac{85}{144}$$

(b) The LCM of 12 and 16 is 48.

$$\frac{7}{12} = \frac{7 \times 4}{12 \times 4} = \frac{28}{48} \qquad\qquad \frac{3}{16} = \frac{3 \times 3}{16 \times 3} = \frac{9}{48}$$

$$\frac{7}{12} + \frac{3}{16} = \frac{28}{48} + \frac{9}{48} = \frac{37}{48}$$

(c) The LCM of 32 and 30 is 480.

$$\frac{10}{32} = \frac{10 \times 15}{32 \times 15} = \frac{150}{480} \qquad\qquad \frac{4}{30} = \frac{4 \times 16}{30 \times 16} = \frac{64}{480}$$

$$\frac{10}{32} + \frac{4}{30} = \frac{150}{480} + \frac{64}{480} = \frac{214}{480} = \frac{107}{240}$$

(d) The LCM of 7 and 5 is 35.

$$\frac{3}{7} = \frac{3 \times 5}{7 \times 5} = \frac{15}{35} \qquad\qquad \frac{4}{5} = \frac{4 \times 7}{5 \times 7} = \frac{28}{35}$$

$$\frac{3}{7} + \frac{4}{5} = \frac{15}{35} + \frac{28}{35} = \frac{43}{35} = 1\frac{8}{35}$$

(e) The LCM of 15 and 9 is 45.

$$\frac{2}{15} = \frac{2 \times 3}{15 \times 3} = \frac{6}{45} \qquad\qquad \frac{5}{9} = \frac{5 \times 5}{9 \times 5} = \frac{25}{45}$$

$$1\frac{2}{15} + \frac{5}{9} = 1 + \frac{2}{15} + \frac{5}{9} = 1 + \frac{6}{45} + \frac{25}{45} = 1 + \frac{31}{45} = 1\frac{31}{45}$$

(f) The LCM of 2, 6, and 4 is 12.

$$\frac{1}{2} = \frac{6}{12} \qquad\qquad \frac{5}{16} = \frac{10}{12} \qquad\qquad \frac{3}{4} = \frac{9}{12}$$

$$\frac{1}{2} + \frac{5}{16} + \frac{3}{4} = \frac{6}{12} + \frac{10}{12} + \frac{9}{12} = \frac{25}{12} = 2\frac{1}{12}$$

(g) The LCM of 12, 9, and 8 is 72.

$$\frac{5}{12} = \frac{5 \times 6}{12 \times 6} = \frac{30}{72} \qquad \frac{1}{9} = \frac{1 \times 8}{9 \times 8} = \frac{8}{72} \qquad \frac{3}{8} = \frac{3 \times 9}{8 \times 9} = \frac{27}{72}$$

$$2\frac{5}{12} + 1\frac{1}{9} + 2\frac{3}{8} = 2 + 1 + 2 + \frac{5}{12} + \frac{1}{9} + \frac{3}{8} = 5 + \frac{30}{72} + \frac{8}{72} + \frac{27}{72} =$$

$$5 + \frac{65}{72} = 5\frac{65}{72}$$

(h) The LCM of 16, 18, and 12 is 144.

$$\frac{13}{16} = \frac{117}{144} \qquad \frac{1}{18} = \frac{8}{144} \qquad \frac{11}{12} = \frac{132}{144}$$

$$\frac{13}{16} + 2\frac{1}{18} + 1\frac{11}{12} = 2 + 1 + \frac{117}{144} + \frac{8}{144} + \frac{132}{144} = 3 + \frac{257}{144} = 4 + \frac{113}{144}$$

$$= 4\frac{113}{144}$$

Once you have mastered the process of adding fractions, subtraction is very simple indeed. Try this problem:

$$\frac{3}{8} - \frac{1}{8} = \underline{\qquad}$$

Turn to **46**.

49

(a) The LCM of 4 and 12 is 12.

$$3\frac{1}{4} = \frac{13}{4} = \frac{13 \times 3}{4 \times 3} = \frac{39}{12} \qquad 1\frac{1}{12} = \frac{13}{12}$$

$$3\frac{1}{4} - 1\frac{1}{12} = \frac{39}{12} - \frac{13}{12} = \frac{26}{12} = \frac{13}{6} = 2\frac{1}{6}$$

(b) $10 = \frac{10}{1} = \frac{10 \times 16}{1 \times 16} = \frac{160}{16} \qquad 3\frac{5}{16} = \frac{53}{16}$

$$10 - 3\frac{5}{16} = \frac{160}{16} - \frac{53}{16} = \frac{107}{16} = 6\frac{11}{16}$$

(c) The LCM of 18 and 12 is 36.

$$4\frac{7}{18} = \frac{79}{18} = \frac{79 \times 2}{18 \times 2} = \frac{158}{36} \qquad 2\frac{11}{12} = \frac{35}{12} = \frac{35 \times 3}{12 \times 3} = \frac{105}{36}$$

$$4\frac{7}{18} - 2\frac{11}{12} = \frac{158}{36} - \frac{105}{36} = \frac{53}{36} = 1\frac{17}{36}$$

(d) The LCM of 32 and 6 is 96.

$$2\frac{6}{32} = \frac{70}{32} = \frac{70 \times 3}{32 \times 3} = \frac{210}{96} \qquad 1\frac{1}{6} = \frac{7}{6} = \frac{7 \times 16}{6 \times 16} = \frac{112}{96}$$

$$2\frac{6}{32} - 1\frac{1}{6} = \frac{210}{96} - \frac{112}{96} = \frac{98}{96} = \frac{49}{48} = 1\frac{1}{48}$$

Now turn to **50** for a set of practice problems on addition and subtraction of fractions.

50

Problem Set 4: Adding and Subtracting Fractions

A. Add or subtract as shown:

$$\frac{3}{5} + \frac{4}{5} = \qquad\qquad \frac{5}{12} + \frac{11}{12} = \qquad\qquad \frac{1}{8} + \frac{7}{8} =$$

$$\frac{7}{15} + \frac{3}{15} = \qquad\qquad \frac{7}{8} - \frac{5}{8} = \qquad\qquad \frac{3}{4} - \frac{1}{4} =$$

$$\frac{7}{12} - \frac{2}{12} = \qquad\qquad \frac{15}{16} - \frac{9}{16} = \qquad\qquad \frac{53}{64} + \frac{19}{64} =$$

$$\frac{17}{32} + \frac{19}{32} = \qquad\qquad \frac{53}{64} - \frac{5}{64} = \qquad\qquad \frac{15}{32} - \frac{7}{32} =$$

$$\frac{2}{16} + \frac{7}{16} + \frac{13}{16} = \qquad\qquad\qquad \frac{7}{18} + \frac{10}{18} + \frac{13}{18} =$$

$$\frac{1}{7} + \frac{6}{7} + \frac{4}{7} = \qquad\qquad\qquad \frac{7}{8} + \frac{1}{8} - \frac{3}{8} =$$

$$\frac{5}{12} + \frac{11}{12} - \frac{1}{12} = \qquad\qquad\qquad \frac{7}{20} - \frac{11}{20} + \frac{13}{20} =$$

B. Add or subtract as shown:

$$\frac{7}{8} + \frac{3}{4} = \qquad\qquad \frac{7}{8} - \frac{3}{4} \qquad\qquad \frac{11}{12} + \frac{7}{18} =$$

$$\frac{11}{12} - \frac{7}{18} = \qquad\qquad \frac{5}{12} + \frac{3}{16} = \qquad\qquad \frac{3}{7} + \frac{2}{5} =$$

$$\frac{11}{48} + \frac{3}{64} = \qquad\qquad \frac{10}{27} + \frac{15}{24} = \qquad\qquad \frac{7}{12} - \frac{5}{16} =$$

$$\frac{5}{7} - \frac{3}{5} = \qquad\qquad \frac{18}{64} - \frac{11}{48} = \qquad\qquad \frac{16}{27} - \frac{5}{24} =$$

$$1\frac{1}{4} + \frac{3}{8} = \qquad\qquad 2\frac{3}{4} + 1\frac{5}{18} = \qquad\qquad 3\frac{5}{12} + 1\frac{15}{16} =$$

$$3\frac{5}{8} + 1\frac{2}{7} = \qquad\qquad 3\frac{5}{8} - 2\frac{7}{8} = \qquad\qquad 3\frac{5}{12} - 1\frac{15}{16} =$$

C. Calculate:

$2 - \dfrac{1}{3} =$ $\qquad\qquad\qquad$ $6 - 4\dfrac{3}{16} =$

$8 - \dfrac{11}{4} =$ $\qquad\qquad\qquad$ $3 - 1\dfrac{1}{5} =$

$6\dfrac{1}{2} + 5\dfrac{3}{4} + 8\dfrac{1}{2} =$ $\qquad\qquad$ $\dfrac{7}{8} - 1\dfrac{1}{3} + 2\dfrac{1}{5} =$

$\dfrac{1}{2} + \dfrac{1}{5} + \dfrac{1}{8} =$ $\qquad\qquad$ $\dfrac{1}{2} + \dfrac{1}{3} + \dfrac{1}{4} + \dfrac{1}{5} =$

$1\dfrac{2}{3}$ subtracted from $4\dfrac{3}{4} =$

$2\dfrac{3}{8}$ less than $4\dfrac{7}{10} =$

$6\dfrac{5}{12}$ reduced by $1\dfrac{3}{16} =$

$4\dfrac{3}{5}$ less than $6\dfrac{1}{2} =$

D. Brain Boosters

1. By how much is $1\dfrac{8}{7}$ larger than $1\dfrac{7}{8}$?

2. Sara's time card shows that she worked the following hours last week: Monday, $7\dfrac{1}{4}$ hours; Tuesday, $6\dfrac{1}{2}$ hours; Wednesday, $5\dfrac{3}{4}$ hours; Thursday, 8 hours; Friday, $9\dfrac{1}{6}$ hours. What total time did she work? What was her total pay at $\$2\dfrac{1}{2}$ per hour?

3. Show that: $(6 + \dfrac{1}{4}) \times (5 - \dfrac{1}{5}) = 6 \times 5$

$\qquad\qquad\qquad$ $(7 + \dfrac{3}{7}) \times (4 - \dfrac{3}{13}) = 7 \times 4$

$\qquad\qquad\qquad$ $(31 + \dfrac{1}{2}) \times (21 - \dfrac{1}{3}) = 31 \times 21$

4. The ancient Egyptians wrote all fractions as a sum of different fractions with numerator 1. For example,

$\qquad\qquad$ $\dfrac{3}{4}$ would be written $\dfrac{1}{2} + \dfrac{1}{4}$

$\dfrac{2}{3}$ would be written $\dfrac{1}{2} + \dfrac{1}{6}$ (not $\dfrac{1}{3} + \dfrac{1}{3}$)

How would the Egyptians have written these fractions?

(a) $\dfrac{7}{8}$ (b) $\dfrac{5}{9}$ (c) $\dfrac{5}{12}$

5. The four sides of a plot of land are $120\dfrac{3}{4}$ ft, $85\dfrac{5}{8}$ ft, $116\dfrac{2}{3}$ ft, and $91\dfrac{5}{6}$ ft. What is the total distance around the edge of this lot?

6. On a recent trip we travelled 1846 miles and stopped for gas three times, using $8\dfrac{1}{2}$ gallons, $10\dfrac{3}{5}$ gallons, and $9\dfrac{3}{10}$ gallons.

(a) How much gas was used?

(b) What was our average miles per gallon?

7. Stock in the Acme Celery Company opened at $47\dfrac{3}{8}$ yesterday on the New York Stalk Exchange and closed at $45\dfrac{3}{16}$. What was its net loss in price?

8. Which is larger?

(a) one-quarter of the sum of $\dfrac{1}{4}$ plus $\dfrac{1}{4}$ of $\dfrac{1}{4}$

(b) $\dfrac{1}{4}$ minus $\dfrac{1}{4}$ of $\dfrac{1}{4}$

The answers to these problems are on pages 228-229. When you have had the practice you need either return to the preview test on page 76 or continue in **51** with the study of word problems involving fractions.

Word Problems

51

The mathematics you use every day in your work or at play usually appears wrapped in words and hidden in sentences. Neat little sets of directions are seldom attached; no "Divide and reduce to lowest terms" or "Write as equivalent fractions and add." The difficulty with real problems is that they must be translated from words to mathematics. You need to learn to <u>talk</u> arithmetic, not just juggle numbers.

Certain words and phrases appear again and again in arithmetic. They are signals alerting you to the mathematical operations to be done. Here is a list of such <u>signal</u> <u>words</u>:

56

$(\frac{3}{4})$ (of) (what number) (is) $(1\frac{2}{3})$

$$\frac{3}{4} \quad \times \quad \square \quad = \quad 1\frac{2}{3} \quad \text{or} \quad \frac{3}{4} \times \square = 1\frac{2}{3}$$

Make up a similar, but very easy, problem:

"Two times what number equals six?"

The answer is 3. How did you get it?

$$\frac{3}{4} \times \square = 1\frac{2}{3}$$

$$\square = 1\frac{2}{3} \div \frac{3}{4}$$

$$\square = \frac{5}{3} \times \frac{4}{3}$$

$$\square = \frac{20}{9}$$

Check: $\dfrac{3}{4} \times \boxed{\dfrac{20}{9}} = \dfrac{5}{3} = 1\dfrac{2}{3}$

Similar Problem

$$2 \times \square = 6$$

$$= 6 \div 2$$

$$= 3$$

Check: $2 \times \boxed{3} = 6$

$$6 = 6$$

If you are one of those people who like to memorize things, it may help you to remember that:

if	$A \times B = C$
then	$B = C \div A$
and	$A = C \div B$

(for any numbers A, B, and C not equal to zero)

As a general rule, it is better not to try to memorize a rule for every variety of math problem and far better to develop your ability to use basic principles, but if you like to memorize helpful rules this is a very useful one.

Use these problems for practice. Translate each into an equation and solve it.

(a) $\dfrac{3}{8}$ of what number is equal to $1\dfrac{5}{16}$?

(b) What fraction of $4\dfrac{1}{2}$ is $6\dfrac{3}{4}$?

(c) Find a number such that $\dfrac{3}{11}$ of it is $2\dfrac{2}{5}$.

(d) The product of $1\dfrac{7}{8}$ and $2\dfrac{1}{3}$ is what number?

(e) What fraction of $8\dfrac{3}{4}$ is $\dfrac{7}{12}$?

The answers are in **58.**

57

   This is similar to the problem "What number times 4 equals 8?"

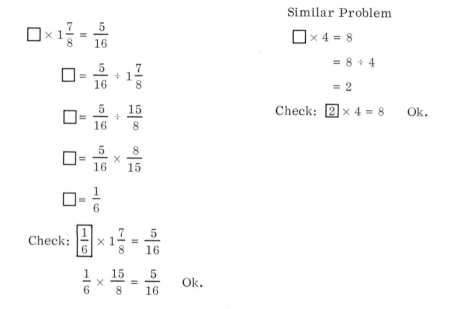

Similar Problem

$$\square \times 1\dfrac{7}{8} = \dfrac{5}{16}$$

$$\square = \dfrac{5}{16} \div 1\dfrac{7}{8}$$

$$\square = \dfrac{5}{16} \div \dfrac{15}{8}$$

$$\square = \dfrac{5}{16} \times \dfrac{8}{15}$$

$$\square = \dfrac{1}{6}$$

$$\square \times 4 = 8$$

$$= 8 \div 4$$

$$= 2$$

Check: $\boxed{2} \times 4 = 8$ Ok.

Check: $\boxed{\dfrac{1}{6}} \times 1\dfrac{7}{8} = \dfrac{5}{16}$

$$\dfrac{1}{6} \times \dfrac{15}{8} = \dfrac{5}{16}$$ Ok.

Isn't it easier when you have the simple problem as a guideline?
   Try another problem:

$$\dfrac{3}{4} \text{ of what number is } 1\dfrac{2}{3} ?$$

Translate this word phrase to an equation in symbols and solve. Check your work in **56.**

A MEMORY GIMMICK

If $A \times B = C$ then $A = \dfrac{C}{B}$ and $B = \dfrac{C}{A}$ for any numbers A, B, and C that are not zero.

Do you need a memory jogger? Try this one.

Examples:

$A \times B = C$

$A = \dfrac{C}{B}$

$B = \dfrac{C}{A}$

$3 \times 4 = 12$

$3 = \dfrac{12}{4}$

$4 = \dfrac{12}{3}$

(a) $(\dfrac{3}{8})$ (of) (what number) (is equal to) $(1\dfrac{5}{16})$

$$\frac{3}{8} \times \square = 1\frac{5}{16}$$

$$\square = 1\frac{5}{16} \div \frac{3}{8}$$

$$\square = \frac{21}{16} \times \frac{8}{3}$$

$$\square = \frac{7}{2} = 3\frac{1}{2} \qquad \text{Check:} \quad \frac{3}{8} \times \boxed{\frac{7}{2}} = \frac{21}{16} = 1\frac{5}{16}$$

(b) (What fraction) (of) $(4\frac{1}{2})$ (is) $(6\frac{3}{4})$?

$$\square \times 4\frac{1}{2} = 6\frac{3}{4}$$

$$\square = 6\frac{3}{4} \div 4\frac{1}{2}$$

$$\square = \frac{27}{4} \times \frac{2}{9}$$

$$\square = \frac{3}{2} = 1\frac{1}{2} \qquad \text{Check: } \boxed{\frac{3}{2}} \times 4\frac{1}{2} = \frac{3}{2} \times \frac{9}{2} = \frac{27}{4} = 6\frac{3}{4}$$

(c) Find (a number) such that $(\frac{3}{11})$ (of) (it) (is) $(2\frac{2}{5})$.

$$\frac{3}{11} \times \square = 2\frac{2}{5}$$

$$\square = 2\frac{2}{5} \div \frac{3}{11}$$

$$\square = \frac{12}{5} \times \frac{11}{3}$$

$$\square = \frac{44}{5} = 8\frac{4}{5} \qquad \text{Check: } \frac{3}{11} \times \boxed{\frac{44}{5}} = \frac{12}{5} = 2\frac{2}{5}$$

(d) The (product of) $(1\frac{7}{8})$ and $(2\frac{1}{3})$ (is) (what number)?

$$1\frac{7}{8} \times 2\frac{1}{3} = \square$$

$$\frac{15}{8} \times \frac{7}{3} = \square$$

$$\frac{35}{8} = \square$$

$$\square = 4\frac{3}{8}$$

(e) (What fraction) (of) $(8\frac{3}{4})$ (is) $(\frac{7}{12})$?

$$\square \times 8\frac{3}{4} = \frac{7}{12}$$

$$\square = \frac{7}{12} \div 8\frac{3}{4}$$

$$\square = \frac{7}{12} \div \frac{35}{4}$$

$$\square = \frac{7}{12} \times \frac{4}{35}$$

$$\square = \frac{1}{15} \qquad \text{Check: } \boxed{\frac{1}{15}} \times 8\frac{3}{4} = \frac{1}{15} \times \frac{35}{4} = \frac{7}{12}$$

Here is another type of problem involving fractions:

If $6\frac{2}{3}$ dozen doorknobs cost \$100, what will 10 dozen door-knobs cost?

To solve this, first make up a similar and much easier problem in order to get the feel of it. For example,

If 2 apples cost 10¢, what will 6 apples cost?

Solution: 2 cost 10¢
  1 costs 10¢ ÷ 2 = 5¢
  then 6 cost 6 × 5¢ = 30¢

In other words we solve the problem by dividing it into two parts:

- First find the unit cost, the cost of one item.
- Second find the cost of any number of items.

Apply this method to the doorknob problem above. Work exactly as you did in the simpler problem. Check your work in **59**.

59

$6\frac{2}{3}$ cost \$100

1 costs $(\$100 \div 6\frac{2}{3}) = \$100 \div \frac{20}{3}$

$$= \$100 \times \frac{3}{20} = \$15$$

Then 10 cost (10 × \$15) = \$150

One way to check your answer is to compare it with what you might have guessed the answer to be before you did the arithmetic. For example, if $6\frac{2}{3}$ cost \$100, then 10 will cost almost double that and the answer will be between \$100 and \$200. The actual answer \$150 is reasonable.

Here are a few problems for you. **Good Luck**

(a) If $4\frac{1}{2}$ dozen pencils cost $2\frac{1}{4}$ dollars, what will $8\frac{1}{2}$ dozen cost?

(b) If you walk 24 miles in $7\frac{1}{2}$ hours, how many miles will you walk in 10 hours, assuming you go at the same rate?

(c) If you can swim $\frac{7}{8}$ of one mile in $\frac{2}{3}$ hour, how far can you swim in $2\frac{1}{2}$ hours at the same rate?

Look in **60** for the answers.

60

(a) $4\frac{1}{2}$ cost $\$2\frac{1}{4}$

1 costs $\$2\frac{1}{4} \div 4\frac{1}{2} = \dfrac{\$9}{4} \div \dfrac{9}{2}$

$\qquad = \dfrac{\$9}{4} \times \dfrac{9}{2} = \dfrac{\$1}{2}$

Then $8\frac{1}{2}$ cost $8\frac{1}{2} \times \dfrac{\$1}{2} = \dfrac{17}{2} \times \dfrac{\$1}{2}$

$\qquad = \dfrac{17}{4} = \$4\frac{1}{4}$

$8\frac{1}{2}$ dozen cost $\$4\frac{1}{4}$

Similar Problem

2 cost 10¢
1 costs 10¢ ÷ 2 = 5¢
6 cost 6 × 5¢ = 30¢

(b) $7\frac{1}{2}$ hours $\longrightarrow$ 24 miles

1 hour $\longrightarrow$ $24 \div 7\frac{1}{2}$

$$= 24 \times \frac{2}{15} = \frac{16}{5} = 3\frac{1}{5} \text{ miles}$$

Then 10 hours $\longrightarrow$ $10 \times \frac{16}{5} = 32$ miles

(c) $\frac{2}{3}$ hour for $\frac{7}{8}$ mile

1 hour $\longrightarrow$ $\frac{7}{8} \div \frac{2}{3} = \frac{7}{8} \times \frac{3}{2} = \frac{21}{16}$ mile

$2\frac{1}{2}$ hours $\longrightarrow$ $\frac{21}{16} \times \frac{5}{2} = \frac{105}{32} = 3\frac{11}{32}$ hours

Now go to **61** for some practice on word problems involving fractions.

61

Problem Set 5: Word Problems

A. Translate these words and phrases to mathematical expressions and symbols:

(a) is

(b) of

(c) increased by

(d) the same as

(e) 6 subtracted from some number

(f) half of a number

(g) double the number

(h) a number divided by $2\frac{1}{2}$

(i) a number plus $\frac{2}{3}$

(j) the sum of a number and $\frac{2}{5}$

(k) a number divided by $\frac{3}{7}$

(l) $\frac{7}{8}$ of a number is equal to $1\frac{1}{2}$

(m) some number less $1\frac{3}{4}$

(n) What fraction of $3\frac{1}{4}$ is $11\frac{1}{2}$?

(o) the product of $\frac{7}{16}$ and $3\frac{5}{8}$

B. Solve:

(a) What fraction of $3\frac{1}{4}$ is $4\frac{7}{8}$?

(b) What part of $7\frac{1}{2}$ is $2\frac{1}{4}$?

(c) What fraction of $3\frac{1}{8}$ is 5?

(d) $\frac{7}{8}$ of what number is $\frac{2}{3}$?

(e) $\frac{4}{11}$ of what number is $\frac{1}{2}$?

(f) $3\frac{1}{8}$ of $\frac{2}{5}$ is what number?

(g) Find a number such that $\frac{5}{16}$ of it is $4\frac{1}{2}$.

C. Solve:

(a) If $4\frac{7}{8}$ pounds of beans cost 39 cents, what will be the cost of $6\frac{1}{2}$ pounds?

(b) A runner travels $6\frac{1}{4}$ miles in 35 minutes. How many miles will he cover in $45\frac{1}{2}$ minutes at this pace?

(c) On a vacation trip, $25\frac{3}{5}$ gallons of gas were used to cover 480 miles. How many gallons were used to cover the first 100 miles at that rate?

(d) If a box containing $2\frac{3}{4}$ pounds of nails cost 50 cents, how many pounds can be purchased for 75 cents?

(e) If you are paid \$138 for $34\frac{1}{2}$ hours of work, what should you be paid for $46\frac{1}{2}$ hours of work at the same rate of pay?

The answers to these problems are on page 229. When you have had the practice you need turn to **62** for a self-test on fractions.

62

Chapter 2 Self-Test

1. Write $7\dfrac{3}{16}$ as an improper fraction: _____

2. Write $\dfrac{37}{11}$ as a mixed number: _____

3. $\dfrac{3}{8} = \dfrac{?}{40}$ _____

4. Reduce to lowest terms: $\dfrac{195}{255}$

5. $\dfrac{3}{5} + \dfrac{2}{7} =$ _____

6. $\dfrac{1}{4} + \dfrac{2}{3} + \dfrac{2}{5} =$ _____

7. $1\dfrac{3}{8} + 2\dfrac{1}{4} + 2\dfrac{2}{3} =$ _____

8. $\dfrac{3}{4} - \dfrac{1}{3} =$ _____

9. $2\dfrac{2}{5} - 1\dfrac{1}{4} =$ _____

10. $6\dfrac{2}{3} - 3\dfrac{1}{4} =$ _____

11. $\dfrac{9}{15} \times \dfrac{5}{3} =$ _____

12. $2\dfrac{2}{7} \times 2\dfrac{1}{4} =$ _____

13. $\dfrac{2}{3} \div \dfrac{3}{5} =$ _____

14. $3\dfrac{2}{7} \div 7\dfrac{1}{3} =$ _____

15. $1\dfrac{3}{16} \div 4\dfrac{3}{4} =$ _____

16. $\left(2\dfrac{1}{3}\right)^2 =$ _____

17. $1\dfrac{3}{5} \times 4\dfrac{7}{8} \times 7\dfrac{1}{2} =$ _____

18. What fraction of 16 is 7? _____

19. What fraction of $7\dfrac{1}{2}$ is 3? _____

20. What fraction of $4\dfrac{2}{3}$ is $3\dfrac{1}{2}$? _____

21. $\dfrac{7}{8}$ of what number is $1\dfrac{3}{4}$? _____

22. $1\dfrac{2}{3}$ of what number is $\dfrac{7}{15}$? _____

23. Find a number such that $\dfrac{2}{7}$ of it is $3\dfrac{1}{2}$. _____

24. What fraction of 50 is $4\frac{1}{2}$? _____

25. If $16\frac{2}{3}$ pounds of peanuts cost 60 cents, what will 20 pounds cost?

_____

The answers to these problems are on page 234-235.

CHAPTER THREE
Decimals

Preview

Objectives	Where to Go for Help	

4. Work with decimal fractions.   Page   Frame

   (a) What decimal part of 16.2 is 4.131?   163   **25**
        (Round to three decimal places.) _____

   (b) If 0.7 of a number is 12.67, find the   163   **25**
        number. _____

   If you are certain you can work all of these problems correctly, turn
to page 174 for a self-test. If you want help with any of these objectives
or if you cannot work one of the preview problems, turn to the page indicated. Super-students eager to learn everything in this unit will turn to
frame **1** and begin work there.

ANSWERS TO PREVIEW PROBLEMS

4. (a) 0.255
    (b) 18.1

3. (a) 0.3714285
    (b) $7\dfrac{13}{40}$

2. (a) 60.03
    (b) 716.856
    (c) 46.875
    (d) 4.16
    (e) 106.09

1. (a) 24.0762
    (b) 1.993

Decimals

1

© 1965 United Feature Syndicate, Inc.

Little Sally has allowed the words to get in the way of what she wants to learn. She may not be able to add 2 and 2 in class or recognize equivalent sets, but you can be certain she makes no mistakes checking her change at the local candy store or adding the money in her piggy bank. Words never interfere with the really important things.

In Chapter 1 you learned that whole numbers are written in a place value system based on powers of ten. A number such as 237 is shown below in expanded form:

$$237$$
$$(2 \times 100) + (3 \times 10) + (7 \times 1)$$

This way of writing numbers can be extended to fractions. A <u>decimal</u> number is a fraction whose denominator is a power of 10. (Remember that a "power of 10" is simply a multiple of 10, that is 10, 100, 1000, 10,000, and so on.) For example,

decimal form			fraction form
.6	=	6 tenths	$= \dfrac{6}{10}$
.05	=	5 hundredths	$= \dfrac{5}{100}$
.32	=	32 hundredths	$= \dfrac{32}{100}$
.004	=	4 thoudandths	$= \dfrac{4}{1000}$
.267	=	267 thousandths	$= \dfrac{267}{1000}$

We may also write the decimal number .267 in expanded form as shown here:

$$.267$$

$$\frac{2}{10} + \frac{6}{100} + \frac{7}{1000}$$

Write the decimal number .526 in expanded form.   Check your answer in **2.**

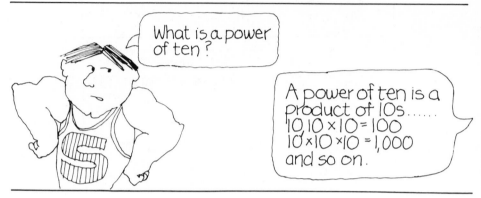

2 $.526 = \dfrac{5}{10} + \dfrac{2}{100} + \dfrac{6}{1000}$

Decimal notation enables us to extend the idea of place value to numbers less than one.   A decimal number often has both a whole number part and a fraction part.   For example, the number 324.576 is shown in expanded form below:

$$324.576$$

$$(3 \times 100) + (2 \times 10) + (4 \times 1) + (5 \times \frac{1}{10}) + (7 \times \frac{1}{100}) + (6 \times \frac{1}{1000})$$

| hundreds | tens | ones | tenths | hundredths | thousandths |

whole number part   fraction part

You are already familiar with this way of interpreting decimal numbers from working with money.

$$\$243.78$$

$$(2 \times \$100) + (4 \times \$10) + (3 \times \$1) + (7 \times 10¢) + (8 \times 1¢)$$

Write the following in expanded form.

(a) $86.42

(b) 43.607

(c) 14.5060

(d) 235.22267

Compare your answer with ours in **3.**

3

(a) $86.42 = (8 \times 10) + (6 \times 1) + (4 \times \frac{1}{10}) + (2 \times \frac{1}{100})$

$= 80 + 6 + \frac{4}{10} + \frac{2}{100}$

(b) $43.607 = (4 \times 10) + (3 \times 1) + (6 \times \frac{1}{10}) + (0 \times \frac{1}{100}) + (7 \times \frac{1}{1000})$

$= 40 + 3 + \frac{6}{10} + \frac{0}{100} + \frac{7}{1000}$

(c) $14.5060 = (1 \times 10) + (4 \times 1) + \frac{5}{10} + \frac{0}{100} + \frac{6}{1000} + \frac{0}{10000}$

(d) $235.22267 = (2 \times 100) + (3 \times 10) + (5 \times 1) + \frac{2}{10} + \frac{2}{100} + \frac{2}{1000} +$

$\frac{6}{10000} + \frac{7}{100000}$

Notice that the denominators in the decimal fractions increase by a factor of 10. For example,

3247 . 8956

3 × 1000	thousands	3000 . 0006	ten-thousandths	6 × 0.0001
2 × 100	hundreds	200 . 005	thousandths	5 × 0.001
4 × 10	tens	40 . 09	hundredths	9 × 0.01
7 × 1	ones	7 . 8	tenths	8 × 0.1

—— Each row changes by a factor of ten. ——

$1 \times 10 = 10$ $0.01 \cdot \times 10 = 0.1$

$10 \times 10 = 100$ $0.001 \times 10 = 0.01$

$100 \times 10 = 1000$ $0.0001 \times 10 = 0.001$

In the decimal number 86.423 the digits 4, 2, and 3 are called deci-mal digits. The number 43.6708 has four decimal digits. All digits to the right of the decimal point, those that name the fractional part of the number, are decimal digits.

How many decimal digits are included in the numeral 324.0576? Count them, then turn to **4**.

4

The number 324.0576 has four decimal digits: 0, 5, 7, and 6 (the digits to the right of the decimal point).

Notice that the decimal point is simply a way of separating the whole number part from the fraction part; it is a place marker. In whole numbers the decimal point usually is not written but its location should be clear to you.

$$2 = 2. \qquad\qquad\qquad 324 = 324.$$

the decimal point   the decimal point

Very often additional zeros are annexed to decimal numbers without changing the value of the original number. For example,

$$8.5 = 8.50 = 8.5000 \quad \text{(and so on)}$$
$$6 = 6. = 6.0 = 6.00 \quad \text{(and so on)}$$

The value of the number is not changed but the additional zeros may be useful, as we shall see.

 The decimal number .6 is often written 0.6. The zero added on the left is used to call attention to the decimal point. It is easy to mistake .6 for 6, but the decimal point in 0.6 cannot be overlooked.

Add the following decimal numbers:

$$0.2 + 0.5 = \frac{2}{10} + \frac{5}{10} = \underline{\hspace{2cm}}$$

Try it, using what you know about adding fractions. Then turn to **5.**

5

$$0.2 + 0.5 = \frac{2}{10} + \frac{5}{10} = \frac{2 + 5}{10} = \frac{7}{10} = 0.7$$

We could skip the second step and say simply: $0.2 + 0.5 = 0.7.$

Because decimal numbers represent fractions with denominators equal to powers of ten, addition is very simple.

$$\begin{aligned}
2.34 \quad &= \quad 2 + \frac{3}{10} + \frac{4}{100} \\
+\,5.23 \quad &= \quad 5 + \frac{2}{10} + \frac{3}{100} \\
\hline
&\quad\;\; 7 + \frac{5}{10} + \frac{7}{100} \quad = \quad 7.57
\end{aligned}$$

Add, using the expanded form shown above:

$$\begin{aligned}
&1.45 \\
+\;&3.42 \\
\hline
\end{aligned}$$

Check your answer in **6.**

6

$$1.45 = 1 + \frac{4}{10} + \frac{5}{100}$$

$$+ 3.42 = 3 + \frac{4}{10} + \frac{2}{100}$$

$$4 + \frac{8}{10} + \frac{7}{100} = 4.87$$

Of course we need not use expanded form in order to add decimal numbers. As with whole numbers, we may arrange the digits in vertical columns and add directly.

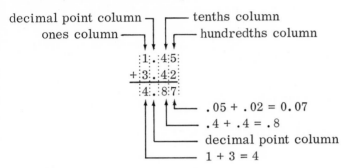

decimal point column — tenths column
    ones column — hundredths column

$$\begin{array}{r} 1.45 \\ + 3.42 \\ \hline 4.87 \end{array}$$

    .05 + .02 = 0.07
    .4 + .4 = .8
    decimal point column
    1 + 3 = 4

Digits of the same power of ten are placed in the same vertical column. Decimal points are always lined up vertically.

If one of the addends is written with fewer decimal digits than the other, annex as many zeros as needed to write both addends with the same number of decimal digits.

$$\begin{array}{r} 2.345 \\ + 1.5 \\ \hline \end{array} \quad \text{becomes} \quad \begin{array}{r} 2.345 \\ + 1.500 \\ \hline \end{array}$$

Except for the preliminary step of lining up decimal points, addition of decimal numbers is exactly the same process as addition of whole numbers.

Add the following decimal numbers.

(a) $4.02 + $3.67 = \underline{\hspace{2cm}}$

(b) $13.2 + 1.57 = \underline{\hspace{2cm}}$

(c) $23.007 + 1.12 = \underline{\hspace{2cm}}$

(d) $14.6 + 1.2 + 3.15 = \underline{\hspace{2cm}}$

(e) $5.7 + 3.4 = \underline{\hspace{2cm}}$

(f) $42.768 + 9.37 = \underline{\hspace{2cm}}$

Arrange each sum vertically, placing the decimal points in the same vertical column. Then add as with whole numbers. Check your work in **7**.

7

(a)
    decimal points in line vertically

$4.02
+ $3.67

7.69 ←— .02 + .07 = .09   add cents
    .0 + .6 = .6   add 10¢ units
    4 + 3 = 7   add dollars

As a check, notice that the sum is roughly $4 + $3 or $7 which agrees with the actual answer. Always check your answer by first estimating it, then comparing your estimate or rough guess with the final answer.

(b)
    decimal points in line

13.20 ←— Annex a zero to provide the same number of decimal
+ 1.57   digits as in the other addend.

14.77
    —— Place answer decimal point in the same vertical line.

Check: 13 + 1 = 14 (which agrees roughly with the answer)

(c)
23.007
+ 1.120

24.127

(d)
14.60
1.20
+ 3.15

18.95

(e)
5.7
+ 3.4

9.1
    —— 7 + .4 = 1.1; write .1, carry /.
    —— carried / + 5 + 3 = 9

In expanded form: $5.7 \quad = \quad 5 + \dfrac{7}{10}$

$$+ \, 3.4 \quad = \quad 3 + \dfrac{4}{10}$$

the carried /

$$8 + \dfrac{7+4}{10} = 8 + \dfrac{11}{10} = 8 + \dfrac{10}{10} + \dfrac{1}{10}$$

$$= 8 + 1 + \dfrac{1}{10} = 9 + \dfrac{1}{10}$$

$$= 9.1$$

the carried /

(f)
$$\begin{array}{r} \overset{\textit{I I I}}{42.768} \\ +\ 9.370 \\ \hline 52.138 \end{array}$$

In expanded form:

$$42 + \frac{7}{10} + \frac{6}{100} + \frac{8}{1000}$$

$$+\ 9 + \frac{3}{10} + \frac{7}{100} + \frac{0}{1000}$$

$$51 + \frac{10}{10} + \frac{13}{100} + \frac{8}{1000} = 51 + 1 + \frac{10}{100} + \frac{3}{100} + \frac{8}{1000}$$

$$= 52 + \frac{1}{10} + \frac{3}{100} + \frac{8}{1000}$$

$$= 52.138$$

$$\frac{10}{10} = 1$$

$$\frac{13}{100} = \frac{10 + 3}{100}$$

$$\frac{10}{100} = \frac{1}{10}$$

𝔅𝔢𝔴𝔞𝔯𝔢

You must line up the decimal points carefully to be certain of getting a correct answer.

Subtraction is equally simple if you are careful to line up decimal points carefully and attach any needed zeros before you begin work.

Example 1: $437.56 – $41

$$\begin{array}{r} \overset{3\ \textit{13}}{\$4\cancel{3}7}.56 \\ -\ \$\ 41.00 \\ \hline \$396.56 \end{array}$$

decimal points in a vertical line
attach zeros (Remember that $41 is $41. or $41.00.)
answer decimal point in the same vertical line

Example 2: 19.452 – 7.3615

$$\begin{array}{r} 19.4\cancel{5}2\cancel{1} \\ -\ 7.3615 \\ \hline 12.0905 \end{array}$$

decimal points in a vertical line
attach zero
answer decimal point in the same vertical line

Try these problems to test yourself on subtraction of decimal numbers.

(a) $37.66 – 14.57 = _____

(b) 248.3 – 135.921 = _____

(c) 6.4701 – 3.2 = _____

(d) 7.304 – 2.59 = _____

Work carefully. The answers are in **8**.

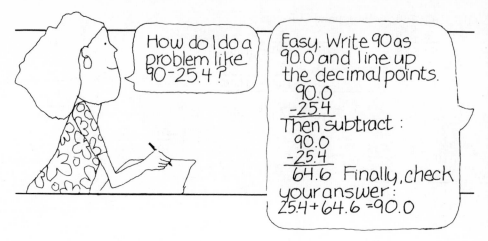

8

(a) $37.66 line up decimal points
     − $14.57
     ‾‾‾‾‾‾‾   Check: 14.57 + 23.09 = 37.66
       23.09

(b) 248.300 line up decimal points
     − 135.921 attach zeros
     ‾‾‾‾‾‾‾
     112.379

              answer decimal point in the same vertical line

        Check: 135.921 + 112.379 = 248.300

(c) 6.4701
     − 3.2000
     ‾‾‾‾‾‾‾
     3.2701   Check: 3.2000 + 3.2701 = 6.4701

(d) 7.304
     − 2.590
     ‾‾‾‾‾‾‾
     4.714   Check: 2.590 + 4.714 = 7.304

Notice that each problem is checked by comparing the sum of the difference (answer) and subtrahend (number subtracted) with the minuend.
Avoid careless mistakes by always checking your answer.

Now for a set of practice problems on addition and subtraction of
decimal numbers turn to **9.**

9

Problem Set 1: Addition and Subtraction of Decimals

A. Add or subtract as shown:

0.5 + 0.3 = _____   0.7 + 0.9 = _____

0.9 + 0.6 = _____   0.4 + 0.2 = _____

0.1 + 0.8 = _____   0.5 + 0.7 = _____

0.8 + 0.8 = _____   0.3 + 0.4 = _____

0.8 + 0.9 = _____   0.8 - 0.7 = _____

0.9 - 0.2 = _____   5.6 - 2.3 = _____

0.9 - 0.4 = _____   4.9 - 2.6 = _____

2.9 - 1.1 = _____   1.7 - 0.3 = _____

3.7 - 0.4 = _____   8.7 - 3.5 = _____

0.7 + 0.9 + 0.3 = _____   5.2 + 1.7 + 3.0 = _____

2.8 + 0.9 + 1.1 = _____   0.5 + 0.8 + 0.1 = _____

2.6 + 4.5 + 1.9 = _____   8.3 + 2.6 + 7.2 = _____

1.4 + 3.6 + 0.5 = _____   0.3 + 0.6 + 0.5 = _____

8.8 + 3.4 + 5.3 = _____   3.3 - 1.7 = _____

9.3 - 2.6 = _____   7.2 - 6.6 = _____

4.2 - 0.6 = _____   7.1 - 5.8 = _____

5.7 - 3.9 = _____   8.5 - 5.9 = _____

3.0 - 0.4 = _____   1.1 - 0.7 = _____

B. Add or subtract as shown:

14.21 + 6.8 = _____ $2.83 + $12.19 = _____

.687 + .93 = _____   3.76 + 23.43 = _____

$7.04 + $23.56 = _____   5.702 + .784 = _____

75.6 + 2.57 = _____ $52.37 + $98.74 = _____

.096 + 5.82 = _____   507.18 + 321.42 = _____

4.0983 + 12.1036 = _____   623.09 + 408.19 = _____

45.6725 + 18.0588 = _____   212.78 + 25.46 = _____

70.3042 + 58.0643 = _____

37 + .09 + 3.5 + 4.605 = _____

$14.75 + $9.08 + $3.76 = _____

.721 + 48.06 + 22 + .09 = _____

$52.19 + $17.43 + $38.75 = _____

24.17 - 4.8 = _____ $33.40 - $18.04 = _____

54.5 - 3.16 = _____   7.83 - 6.79 = _____

$11.36 - $7.50 = _____   75.08 - 32.75 = _____

10.05 - 3.42 = _____ $20.00 - $13.48 = _____

14.22 - 7.8 = _____ $40 - $3.82 = _____

30 - 7.984 = _____   3.892 - .995 = _____

$65 - $47.35 = _____   13 - 6.04 = _____

1.0487 - .6728 = _____

C. Calculate:

148.002 + 3.459 =   632.9 - 30.246 =

68.708 + 27.18 =   517.03 - 425.88 =

7.865 + 308.9 =   23.745 - 9.06745 =

.9437 + 15.0988 =   4,068.4 - 32.9067 =

8.939 + 10.072 =   9.77803 - 6.42829 =

D. Brain Boosters

1. On a recent shopping spree at the Happy Peanut Health Food
    Store you bought the following:

1 qt celery juice	$ .75
1 jar honey	1.46
Granny's Granola	1.38
sunflower seeds	.59
wheat germ oil	3.98

    How much change should you receive from a $10 bill.

2. At the start of a long trip your mileage meter read 18327.4 and
    at the end of the trip it read 23015.2. How far did you travel?

3. The number "nine thousand, nine hundred, nine dollars" is
    written $9,909. Quickly now, write the number "twelve thousand,
    twelve hundred, twelve dollars."

4. The 400 meter race run in the Olympic games is 437.444 yards long. What is the difference between this distance and one-quarter of a mile (440 yards)?

5. A certain machine part is 2.345 inches thick. What is its thickness after 0.078 inches are ground off?

6. One minute is defined as exactly 60 seconds in sun time. One minute as measured by movement of the earth with respect to the stars is 59.836174 seconds. Find the difference between sun time and star time for one minute.

7. Can you balance a checkbook? At the start of a shopping spree your balance was $472.33. While shopping you wrote checks for $12.57, $8.95, $4, $7.48, and $23.98. What is your new balance?

8. Find the perimeter (distance around) the field shown. All distances are in meters.

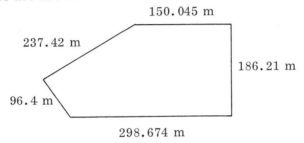

The answers to these problems are on page 230. When you have had the practice you need either return to the preview on page 135 or continue in frame **10** with the study of multiplication and division of decimal numbers.

Multiplication and Division

10

A decimal number is a fraction with a power of ten as denominator. Multiplication of decimals should therefore be no more difficult than the multiplication of fractions. Try this problem:

$$0.5 \times 0.3 = \underline{\hspace{2cm}}$$

Write out the two numbers as fractions and multiply, then choose an answer.

(a) 15   Go to **11.**

(continued)

(b) 1.5   Go to **12**.

(c) .15   Go to **13**.

11

You answered that $0.5 \times 0.3 = 15$ and that is incorrect. A wise first step would be to guess at the answer. Both 0.5 and 0.3 are less than 1, therefore their product is also less than 1, and 15 is not a reasonable answer.

Try calculating the sum this way:

$$0.5 = \frac{5}{10} \qquad 0.3 = \frac{3}{10}$$

$$0.5 \times 0.3 = \frac{5}{10} \times \frac{3}{10}$$

Complete this multiplication, then return to **10** and choose a better answer.

12

Your answer is incorrect. Don't get discouraged; we'll never tell.

The first step is to guess: both 0.5 and 0.3 are less than 1, therefore their product will be less than 1. Next, convert the decimals to fractions with 10 as denominator.

$$0.5 = \frac{5}{10} \qquad 0.3 = \frac{3}{10}$$

Finally, multiply.

$$0.5 \times 0.3 = \frac{5}{10} \times \frac{3}{10} = \underline{\hspace{2cm}}$$

Complete this multiplication, then return to **10** and choose a better answer.

13

Excellent. Notice that both 0.5 and 0.3 are less than 1, therefore their product will be less than 1. This provides a rough guess at the answer.

$$0.5 = \frac{5}{10} \qquad 0.3 = \frac{3}{10}$$

$$0.5 \times 0.3 = \frac{5}{10} \times \frac{3}{10} = \frac{5 \times 3}{10 \times 10} = \frac{15}{100} = 0.15$$

Of course it would be very, very clumsy and time consuming to calculate every decimal multiplication in this way. We need a simpler method. Here is the procedure most often used:

■Step 1: Multiply the two decimal numbers as if they were whole numbers. Pay no attention to the decimal points.

■Step 2: The sum of the decimal digits in the factors will give you the number of decimal digits in the product.

Let's apply this procedure to the problem $3.2 \times .41$.

■Step 1: Multiply without regard to the decimal points.

$$
\begin{array}{r}
32 \\
\times\,41 \\
\hline
1312
\end{array}
$$

■Step 2: Count the number of decimal digits in the factors.

3.2 has <u>one</u> decimal digit (2).
.41 has <u>two</u> decimal digits (4 and 1).

The total number of decimal digits in the two factors is 3. The product will have <u>three</u> decimal digits. Count over <u>three</u> digits to the left in the product.

1.312

three decimal digits

Check: $3.2 \times .41$ is roughly $3 \times \dfrac{1}{2}$ or about $1\dfrac{1}{2}$.

The answer 1.3 agrees with our rough guess.

Try these simple decimal multiplications.

(a) $0.5 \times 0.5 =$ _____

(b) $0.1 \times 0.1 =$ _____

(c) $2 \times 0.6 =$ _____

(d) $2 \times 0.4 =$ _____

(e) $1 \times 0.1 =$ _____

(f) $2 \times 0.003 =$ _____

(g) $0.01 \times 0.02 =$ _____

(h) $0.04 \times 0.005 =$ _____

Follow the steps outlined above. Count decimal digits carefully. Check your answers in **14.**

HOW TO NAME DECIMAL NUMBERS

The decimal number 3,254,935.4728 should be interpreted as:

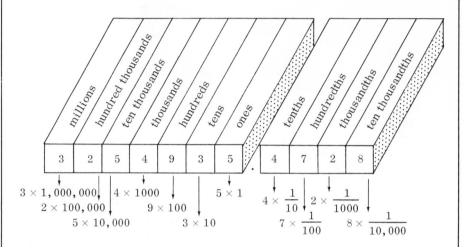

It may be read "three million, two hundred fifty-four thousand, nine hundred thirty-five, and four thousand one hundred twenty-eight ten thousandths."

Notice that the decimal point is read "and."

It is useful to recognize that in this number the digit 8 represents 8 ten-thousandths or $\dfrac{8}{10,000}$ and the digit 7 represents 7 hundredths or $\dfrac{7}{100}$.

This number is often read more simply as "three million, two hundred fifty-four thousand, nine hundred thirty-five, <u>point</u> four, seven, two, eight." This way of reading the number is easiest to write, to say, and to understand.

14

(a) $0.5 \times 0.5 = $ _____

First, multiply $5 \times 5 = 25$. Second, count decimal digits:

0.5 (one decimal digit) $\times$ 0.5 (one decimal digit) =
a total of <u>two</u> decimal digits

Count over <u>two</u> decimal digits from the right: .25
The product is 0.25.

(a) continued

Check: Both factors (0.5) are less than 1, therefore their product will be less than 1, and 0.25 seems reasonable.

(b) 0.1×0.1 $1 \times 1 = 1$

Count over <u>two</u> decimal digits from the right. Since there are not two decimal digits in the product attach a few zeros on the left.

$$1 \longrightarrow 0.01$$

two decimal digits

So $0.1 \times 0.1 = 0.01$

Check: $\dfrac{1}{10} \times \dfrac{1}{10} = \dfrac{1}{100}$ Ok.

(c) 10×0.6 $10 \times 6 = 60$

Count over <u>one</u> decimal digit from the right (6.0) so that $10 \times 0.6 = 6.0$. Notice that multiplication by 10 simply shifts the decimal place one digit to the right.

$$10 \times 6.2 = 62$$
$$10 \times 0.075 = 0.75$$
$$10 \times 8.123 = 81.23 \quad \text{(and so on)}$$

(d) 2×0.4 $2 \times 4 = 8$

Count over <u>one</u> decimal digit.   .8 $2 \times 0.4 = 0.8$

(e) 1×0.1 $1 \times 1 = 1$

Count over <u>one</u> decimal digit.   .1 $1 \times 0.1 = 0.1$

(f) 2×0.003 $2 \times 3 = 6$

Count over <u>three</u> decimal digits.   0.006 $2 \times 0.003 = 0.006$

(g) 0.01×0.02 $1 \times 2 = 2$

0.01 (two decimal digits) $\times$ 0.02 (two decimal digits) = total of four decimal digits

Count over four decimal digits.   0.0002 $0.01 \times 0.02 = 0.0002$

(h) 0.04×0.005 $4 \times 5 = 20$

0.04 (two decimal digits) $\times$ 0.005 (three decimal digits) = total of five decimal digits

Count over five decimal digits.   0.00020 $0.04 \times 0.005 = 0.0002$

𝕽emember . . .

◊ Do not try to do this entire process mentally until you are certain you will not misplace zeros.

◊ Always estimate before you begin the arithmetic, and finally check your answer against your estimate.

Multiplication of larger decimal numbers is performed in exactly the same manner. Try these:

(a) $4.302 \times 12.05 =$ _____

(b) $6.715 \times 2.002 =$ _____

(c) $3.144 \times 0.00125 =$ _____

Look in **15** for the answers.

15

(a) Guess: 4×12 is 48. The product will be about 48.

$$
\begin{array}{r}
4302 \\
\times\ 1205 \\
\hline
5183910
\end{array}
$$

(If you cannot do this multiplication correctly turn to page 29 in Chapter 1 for help with the multiplication of whole numbers.)

The factors contain a total of five decimal digits (three in 4.302 and two in 12.05). Count over five decimal digits from the right in the product

51.83910

so that $4.302 \times 12.05 = 51.8391$

Check: The answer, 51.8391, is approximately equal to the guess, 48.

(b) Guess: 6.7×2 is about 7×2 or 14.

$$
\begin{array}{r}
6715 \\
\times\ 2002 \\
\hline
13.443430
\end{array}
$$

6.715 has <u>three</u> decimal digits

2.002 has <u>three</u> decimal digits

a total of <u>six</u> decimal digits

six decimal digits $6.715 \times 2.002 = 13.44343$

Check: The answer and the guess are approximately equal.

(c) Guess: 3×0.001 is about 0.003.

$$
\begin{array}{r}
3144 \\
\times\ 125 \\
\hline
.00393000
\end{array}
$$

3.144 has <u>three</u> decimal digits

0.00125 has <u>five</u> decimal digits

a total of <u>eight</u> decimal digits

eight decimal digits $3.144 \times 0.00125 = 0.00393$

Check: The answer and the guess are approximately equal.

Now go to **16** for a look at division of decimal numbers.

16

<u>Division</u> of decimal numbers is very similar to division of whole numbers. For example, $6.8 \div 1.7$ can be written:

$$\frac{6.8}{1.7}$$

and if we multiply top and bottom of the fraction by 10,

$$\frac{6.8}{1.7} = \frac{6.8 \times 10}{1.7 \times 10} = \frac{68}{17}$$

$\frac{68}{17}$ is a normal whole number division:

$$\frac{68}{17} = 68 \div 17 = 4$$

Therefore, $6.8 \div 1.7 = 4$.   Check: $1.7 \times 4 = 6.8$.

Rather than take the trouble to write the division as a fraction, we may use a short cut.

Example

■Step 1: Write the divisor and dividend in standard long division form.

$6.8 \div 1.7$

$1.7\overline{)6.8}$

■Step 2: Shift the decimal point in the divisor to the right so as to make the divisor a whole number.

$1.7\overline{)}$

■ Step 3: Shift the decimal point in the dividend the <u>same</u> <u>amount</u>. (Add zeros if necessary.)

$$1.\underset{\smile}{7.}\overline{)6.\underset{\smile}{8.}}$$

■ Step 4: Place the decimal point in the answer space directly above the new decimal position in the dividend.

$$17.\overset{\cdot}{\overline{)68.}}$$

■ Step 5: Complete the division exactly as you would with whole numbers. The decimal points in divisor and dividend may now be ignored.

$$\begin{array}{r} 4. \\ 17.\overline{)68.} \\ \underline{68} \end{array}$$

$$6.8 \div 1.7 = 4$$

Notice in steps 2 and 3 we have simply multiplied both divisor and dividend by 10.

Repeat the process outlined above with this division:

$$1.38 \div 2.3$$

Work carefully, then compare your work with ours in **17.**

17

Let's do it step by step.

$$2.3\overline{)1.38}$$

$$2.\underset{\smile}{3.}\overline{)1.\underset{\smile}{3.}8}$$

Shift the decimal point one digit to the right so that the divisor becomes a whole number. Then shift the decimal point in the dividend the <u>same</u> number of digits. 2.3 becomes 23. 1.38 becomes 13.8. This is the same as multiplying both numbers by 10.

$$23.\overset{\cdot}{\overline{)13.8}}$$

Place the answer decimal point directly above the decimal point in the dividend.

$$\begin{array}{r} .6 \\ 23.\overline{)13.8} \\ \underline{13\ 8} \end{array}$$

Divide as you would with whole numbers.

$$1.38 \div 2.3 = 0.6$$

Check: $2.3 \times 0.6 = 1.38$

Always remember to check your answer.

How would you do this one?

$$2.6 \div 0.052 = \underline{\hspace{2cm}}$$

Look in **18** for the solution after you have tried it.

18

$$.0\underarrow{52.}\,\overline{)2.6}$$

$$.0\underarrow{52.}\,\overline{)2.6\underarrow{00.}}$$

$$\begin{array}{r} 50. \\ 52.\overline{)2600.} \\ \underline{260} \\ 0 \\ \underline{0} \end{array}$$

To shift the decimal place three digits in the dividend we must attach several zeros to its right.

Then place the decimal point in the answer space above that in the dividend, and divide as with whole numbers.

$2.6 \div 0.052 = 50$

Check: $0.052 \times 50 = 2.6$

Shifting the decimal point three digits and attaching zeros to the right of the decimal point in this way is equivalent to multiplying both divisor and dividend by 1000.

Try these problems:

(a) $3.5 \div 0.001 =$ _____

(b) $9 \div 0.02 =$ _____

(c) $.365 \div 18.25 =$ _____

(d) $8.8 \div 3.2 =$ _____

The answers are in **19.**

19

(a) $0.0\underarrow{01.}\,\overline{)3.5\underarrow{00.}}$

$$\begin{array}{r} 3500. \\ 1.\overline{)3500.} \end{array}$$

$3.5 \div 0.001 = 3500$

Check: $0.001 \times 3500 = 3.5$

(b) $0.02\overline{)9.00}$

$$\begin{array}{r} 450. \\ \hline 2.\ \ 900. \end{array}$$

$9 \div 0.02 = 450$

Check: $0.02 \times 450 = 9$

(c) $18.25\overline{).3615}$

$$\begin{array}{r} .02 \\ \hline 1825.\ \ 36.50 \\ 36\ 50 \end{array}$$

$0.365 \div 18.25 = 0.2$

Check: $18.25 \times 0.2 = .365$

(d) $3.2\overline{)8.8}$

$$\begin{array}{r} 2.75 \\ \hline 32.\ \ 88.00 \\ 64\ \ \ \ \ \ \\ 240\ \ \ \\ 224\ \ \ \\ \hline 160 \\ 160 \end{array}$$

$8.8 \div 3.2 = 2.75$

Check: $3.2 \times 2.75 = 8.8$

If the dividend is not exactly divisible by the divisor we must either stop the process after some pre-set number of decimal places in the answer or we must round the answer. We do not generally indicate a remainder in decimal division.

Turn to **20** for some rules for rounding.

___

20

<u>Rounding</u> is a process of approximating a number. To round a number means to find another number roughly equal to the given number but expressed less precisely. For example,

$432.57 = 400$ rounded to the nearest hundred dollars
$\quad\quad\quad\ = 430$ rounded to the nearest ten dollars
$\quad\quad\quad\ = 433$ rounded to the nearest dollar

$1.376521 = 1.377$ rounded to three decimal digits
$\quad\quad\quad\quad\ = 1.4$ rounded to the nearest tenth
$\quad\quad\quad\quad\ = 1$ rounded to the nearest whole number

There are exactly 5280 feet in 1 mile. To the nearest thousand feet how many feet are in one mile? To the nearest hundred feet?

Check your answers in **21.**

21

5280 ft = 5000 ft rounded to the nearest thousand feet
(In other words 5280 is closer to 5000 than to 4000 or 6000.)
5280 ft = 5300 ft rounded to the nearest hundred feet
(In other words 5280 is closer to 5300 than to 5200 or 5400.)

For most rounding follow these simple steps:

Example

■ Step 1: Determine the number of digits or the place to which the number is to be rounded. Mark it with a ∧.

Round 3.462 to one decimal place:
3.4∧62

■ Step 2: If the digit to the right of the mark is less than 5, replace all digits to the right of the mark by zeros. If the zeros are decimal digits you may discard them.

2.8∧32 becomes 2.800 or 2.8

■ Step 3: If the digit to the right of the mark is equal to or larger than 5, increase the digit to the left by 1.

3.4∧62 becomes 3.5

Try applying this rounding procedure to these problems.

(a) Round 74.238 to two decimal places.

(b) Round 8.043 to two decimal places.

(c) Round 156 to the nearest hundred.

(d) Round 6.07 to the nearest tenth.

Follow the rules, then check your work in **22**.

The human brain is a fantastic machine, Isn't it?

Sure. It starts working the moment you're born and never stops—until you pick up a math book.

22

(a) 74.238 = 74.24 to two decimal places
(Write 74.23̰8 and note that 8 is larger than 5 so increase the 3 to 4.)

(b) 8.043 = 8.04 to two decimal places
(Write 8.04̰3 and note that 3 is less than 5 so drop it.)

(c) 156 = 200 to the nearest hundred
(Write 1̰56 and note that the digit to the right of the mark is 5 so increase the 1 to 2.)

(d) 6.07 = 6.1 to the nearest tenth
(Write 6.0̰7 and note that 7 is greater than 5 so increase the 0 to 1.)

There are a few very specialized situations where this rounding rule is not used:

Engineers use a more complex rule when rounding a number that ends in 5.

In business, fractions of a cent are usually rounded up in determining selling price. For example, if three items cost 25 cents, one costs $8\frac{1}{3}$ cents which is rounded to 9 cents.

Our rule will be quite satisfactory for most of your work in arithmetic.
In this problem, divide as shown and round your answer to two decimal places.

6.84 ÷ 32.7 = _____

Careful now. Check your work in **23**.

MULTIPLYING AND DIVIDING BY POWERS OF TEN

Many practical problems involve multiplying or dividing by 10, 100, or 1000. You will find it very useful to be able to multiply and divide by powers of 10 quickly and without using paper and pencil. The following rules will help.

1. To multiply a whole number by 10, 100, 1000, or a larger multiple of 10, attach as many zeros to the right of the number as there are in the multiple of 10.

24 × 10 = 240   24 × 100 = 2400
      attach 1 zero   attach 2 zeros

2. To multiply a decimal number by 10, 100, 1000, or a larger multiple of 10, move the decimal point as many places to the right as there are zeros in the multiplier.

$$3.46 \times 10 = 34.6 \qquad\qquad 3.46 \times 100 = 346.$$

move the decimal point   move the decimal point
1 place to the right   2 places to the right

You may need to attach additional zeros before moving the decimal point.

$$2.4 \times 1000 = 2.400 \times 1000 = 2400.$$

move the decimal point
3 places to the right

3. To divide a whole number or a decimal number by 10, 100, 1000, or a larger multiple of 10, move the decimal point as many places to the left as there are zeros in the divisor.

$$12.4 \div 10 = 1.24 \qquad\qquad 12.4 \div 100 = .124$$

move the decimal point   move the decimal point
1 place to the left   2 places to the left

You may need to attach additional zeros before moving the decimal point.

$$3.4 \div 100 = 03.4 \div 100 = .034$$

move the decimal point
2 places to the left

With a whole number the decimal point is usually not written and you must remember that it is understood to be after the units digit.

$$4 = 4.$$
$$4 \div 100 = 4. \div 100 = .04$$

move the decimal point
2 places to the left

Here are a few problems for practice. Work quickly. No pencil or paper needed. Do them in your head.

1. $4 \times 10 =$ _____   2. $64 \times 100 =$ _____

3. $16 \times 10,000 =$ _____   4. $3.5 \times 1000 =$ _____

5. $1.26 \times 10 =$ _____   6. $4.23 \times 100 =$ _____

7. $.4 \times 10 =$ _____   8. $.004 \times 100 =$ _____

9. $.075 \times 10 = $ _____   10. $.075 \times 1000 = $ _____

11. $1.275 \times 100 = $ _____   12. $2 \times 10,000 = $ _____

13. $45 \div 10 = $ _____   14. $376 \div 100 = $ _____

15. $82.1 \div 10 = $ _____   16. $82.1 \div 100 = $ _____

17. $82.1 \div 10,000 = $ _____   18. $4 \div 1000 = $ _____

19. $0.24 \div 10 = $ _____   20. $6 \div 100 = $ _____

21. $.035 \div 10 = $ _____

23

$$32.7 \overbrace{} \; 6.8\overbrace{}4$$

$$\begin{array}{r} .209 \\ \hline 327.\overline{\smash{)}68.400} \\ 65\ 4\downarrow\downarrow \qquad 2 \times 327 = 654 \\ \hline 3\ 000 \\ 2\ 943 \qquad 9 \times 327 = 2943 \\ \hline \end{array}$$

$0.209 = 0.21$ rounded to two decimal places

$6.84 \div 32.7 = 0.21$ rounded to two decimal places

Check: $32.7 \times 0.21 = 6.867$ which is approximately equal to 6.84.   (The check will not be exact because we have rounded.)

Go to **24** for a set of practice problems on multiplication and division of decimal numbers.

24

Problem Set 2: Multiplication and Division of Decimals

A. Multiply:

$0.01 \times 0.001 = $ $10 \times 0.01 = $

$10 \times 2.15 = $ $3 \times 0.02 = $

$0.04 \times 0.2 = $ $0.07 \times 0.2 = $

$0.3 \times 0.3 = $ $0.9 \times 0.8 = $

$1.2 \times 0.7 = $ $4.5 \times 0.002 = $

$0.005 \times 0.012 = $ $3.5 \times 1.2 = $

6.41 × .23 =

16.2 × 0.031 =

0.5 × 1.2 × 0.04 =

1.2 × 1.23 × 0.01 =

1.2 × 10 × .12 =

0.234 × 0.005 =

5.224 × 0.00625 =

425.6 × 2.875 =

7.25 × .301 =

0.2 × 0.3 × 0.5 =

0.6 × 0.6 × 6.0 =

2.3 × 1.5 × 1.05 =

321.4 × .25 =

125 × 2.3 =

0.1234 × 0.0075 =

B. Divide:

6.5 ÷ 0.005 =

.0405 ÷ 0.9 =

0.378 ÷ 0.003 =

3 ÷ 0.05 =

10 ÷ 0.001 =

1.2321 ÷ 0.11 =

57.57 ÷ 0.0303 =

1.111 ÷ 10.1 =

3.78 ÷ 0.30 =

6.5 ÷ 0.5 =

40.5 ÷ 0.09 =

12 ÷ 0.006 =

2.59 ÷ 70 =

44.22 ÷ 6.7 =

104.2 ÷ .0320 =

C. Divide and round as indicated:

(a) round to two decimal digits

10 ÷ 3 =

5 ÷ 6 =

2.0 ÷ .19 =

3 ÷ .08 =

.023 ÷ .19 =

12.3 ÷ 4.7 =

2.37 ÷ .07 =

6.5 ÷ 1.3 =

1 ÷ 0.7 =

.07 ÷ .80 =

2 ÷ 3 =

.17 ÷ .19 =

345 ÷ 4.7 =

0.16 ÷ 1.35 =

4.27 ÷ .009 =

(b) round to the nearest tenth

100 ÷ 3 =

21.23 ÷ 98.7 =

16 ÷ 15 =

20 ÷ 3 =

$1 \div 4 =$ $1 \div 8 =$

$100 \div 9 =$ $20 \div 0.07 =$

$0.006 \div .04 =$ $.8 \div .05 =$

(c) round to three decimal digits

$10 \div .70 =$ $.04 \div 1.71 =$

$.09 \div 0.40 =$ $0.091 \div .0014 =$

$22.4 \div 6.47 =$ $3.41 \div 0.25 =$

$3.51 \div 0.92 =$ $6.001 \div 2.001 =$

$4.0 \div .007 =$ $123.21 \div 0.1111$

D. Calculate as indicated:

$(0.3)^2 =$ $(0.03)^2 =$ $(0.003)^2 =$

$(0.3)^3 =$ $(1.2)^2 =$ $(1.2)^3 =$

$(0.1)^2 =$ $(0.01)^2 =$ $(0.03)^3 =$

$\dfrac{1}{81}$ (round to 3 decimal digits) =

$\dfrac{1}{7}$ (round to 3 decimal digits) =

$\dfrac{0.23 \times 7.5}{0.23 + 7.5} =$ $\dfrac{.02 \times 3.2}{0.2 + 3.2} =$

$\dfrac{0.065 - 0.042}{0.065 + 0.042} =$ $\dfrac{0.03 \div 0.006}{0.03 \times 0.006} =$

E. Brain Boosters

1. Andy worked 37.4 hours at $2.25 per hour. How much money did he earn?

2. What is the cost of 12.3 gallons of gasoline at 63.9 cents per gallon?

3. A color television set is advertised for $420. It can also be bought "on time" for 24 payments of $22.75 each. How much extra do you pay by purchasing it on the installment plan? (These are actual numbers from a local furniture bandit.)

4. The telephone rates between Santa Barbara and Zanzibar are $10.75 for the first three minutes and $1.95 for each additional minute. What would be the cost of an eleven minute telephone call?

5. If a jet plane averaged 624.38 mph for a 4.6 hour trip, what distance did it travel?

6. There are three errors in these five problems. What are they?

   (a) $6 \div 0.02 = 300$ (b) $12 \div 0.03 = 4$

   (c) $2.4 \times 0.3 = .72$ (d) $100 \times 0.002 = 0.20$

   (e) $0.2 \times 0.2 = 0.4$

7. Lyn once said, "The day before yesterday I was 19, but next year I'll be 22." She always tells the truth. When did she say this?

8. Is this problem correct?

$$\frac{6 + 6 - 0.6}{0.6} = 19$$

Try substituting any other digit for 6. Notice anything interesting?

The answers to these problems are on page 230. When you have had the practice you need either return to the preview on page 135 or continue in frame **25** with the study of decimal fractions.

Decimal Fractions

25

Since decimal numbers are fractions, they may be used, as fractions are used, to represent a part of some quantity. For example, recall that

$$"\frac{1}{2} \text{ of 8 equals 4" means } \frac{1}{2} \times 8 = 4$$

and therefore,

$$"0.5 \text{ of 8 equals 4" means } 0.5 \times 8 = 4$$

The word of used in this way indicates multiplication.

Find 0.35 of 8.4.

(If you need a review of problems of this kind, turn to frame **51**, page 120.) Go to **26** to check your answer to this problem.

26

$0.35 \times 8.4 =$ _____

$$8.4 \leftarrow \text{one decimal digit}$$
$$\underline{\times\, 0.35} \leftarrow \text{two decimal digits}$$
$$420$$
$$\underline{252}$$
$$2.940 \leftarrow \text{a total of three decimal digits}$$

Here is a second variety of problem:

What part of 16 is 4?

$$\square \times 16 = 4$$

$$\square = 4 \div 16 = \frac{1}{4}$$

What decimal part of 16 is 4?

$$\square \times 16 = 4$$

$$\square = 4 \div 16$$

Dividing:
$$\begin{array}{r} .25 \\ 16\overline{)4.00} \\ \underline{3\ 2} \\ 80 \\ \underline{80} \end{array} \qquad \square = 0.25$$

Check: $0.25 \times 16 = 4.00$ or 4

What decimal part of 8 is 3? Solve this problem using the method illustrated above, then hop to **28.**

27

(a) $\square \times 5 = 13$

$\square = 13 \div 5$

$\square = 2.6$

Check: $2.6 \times 5 = 13.0$ or 13

$$\begin{array}{r} 2.6 \\ 5\overline{)13.0} \\ \underline{10} \\ 30 \\ \underline{30} \end{array}$$

(b) $0.8 \times \square = 10$

$\square = 10 \div 0.8$

$\square = 12.5$

Check: $0.8 \times 12.5 = 10.00$ or 10

$$\begin{array}{r} 1\ 2.5 \\ 0.8\overline{)10.0\ 0} \\ \underline{8} \\ 2\ 0 \\ \underline{1\ 6} \\ 4\ 0 \\ \underline{4\ 0} \end{array}$$

(c) $2.35 \times \square = 1.739$

$\square = 1.739 \div 2.35$

$\square = 0.74$

Check: $2.35 \times 0.74 = 1.7390$ or 1.739

$$2.35. \overline{)1.73.90} \quad \begin{array}{r} .74 \\ \hline \end{array}$$
$$\begin{array}{r} 1\ 64\ 5 \\ \hline 9\ 40 \\ 9\ 40 \end{array}$$

To convert a number from fraction form to decimal form simply divide as indicated in the problems above. If the division has no remainder, the decimal number is called a <u>terminating</u> <u>decimal</u>.

For example, $\dfrac{5}{8}$ = _____

$$\begin{array}{r} .625 \\ 8\overline{)5.000} \end{array} \longleftarrow \text{ Attach as many zeros as needed.}$$
$$\begin{array}{r} 4\ 8 \\ \hline 20 \\ 16 \\ \hline 40 \\ 40 \\ \hline \end{array}$$

$\dfrac{5}{8} = 0.625$ ⟵ Zero remainder, hence the decimal terminates or ends.

If a decimal does not terminate, you may round it to any desired number of decimal digits. Try this one.

$\dfrac{2}{13}$ = _____

Divide it out and round to three decimal digits. Check your work in **29.**

28

What decimal part of 8 is 3?

$\square \times 8 = 3$

$\square = 3 \div 8$

$\square = 0.375$

Check: $0.375 \times 8 = 3.000$ or 3

$$\begin{array}{r} .375 \\ 8\overline{)3.000} \end{array}$$
$$\begin{array}{r} 2\ 4 \\ \hline 60 \\ 56 \\ \hline 40 \\ 40 \\ \hline \end{array}$$

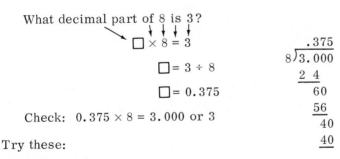

Try these:

(a) What decimal part of 5 is 13?

(b) If 0.8 of a number is 10, find the number.

(c) If 2.35 of a number is 1.739, find the number.

Look in **27** for the answers.

29

$$13\overline{)\begin{array}{r} .1538 \\ 2.0000 \end{array}}$$

$$\begin{array}{r} 1\ 3 \\ \hline 70 \\ 65 \\ \hline 50 \\ 39 \\ \hline 110 \\ 104 \\ \hline 6 \end{array}$$

$\dfrac{2}{13} = 0.154$ rounded to three decimal digits

Convert the following fractions to decimal form and round to two decimal digits.

(a) $\dfrac{2}{3} =$

(b) $\dfrac{5}{6} =$

(c) $\dfrac{17}{7} =$

(d) $\dfrac{7}{16} =$

Our work is in **30.**

30

(a) $3\overline{)\begin{array}{r} .666 \\ 2.000 \end{array}}$

$$\begin{array}{r} 1\ 8 \\ \hline 20 \\ 18 \\ \hline 20 \\ 18 \\ \hline 2 \end{array}$$

$\dfrac{2}{3} = 0.67$ rounded to two decimal digits

Notice that in order to round to two decimal digits we must carry the division out to at least three decimal digits.

(b)
```
     .833
  6)5.000
   4 8
     20
     18
     20
     18
      2
```
$\dfrac{5}{6} = 0.83$ rounded to two decimal digits

(c)
```
    2.428
  7)17.000
   14
    30
    28
    20
    14
    60
    56
     4
```
$\dfrac{17}{7} = 2.43$ rounded to two decimal digits

(d)
```
     .437
 16)7.000
   6 4
    60
    48
   120
   112
     8
```
$\dfrac{7}{16} = 0.44$ rounded to two decimal digits

Decimal numbers that do not terminate repeat a sequence of digits. Such decimals are called <u>repeating</u> decimals. For example,

$$\frac{1}{3} = 0.333 \ . \ . \ .$$

The three dots are read "and so on" and indicate that the digit 3 continues without end. Similarly,

$$\frac{2}{3} = 0.6666 \ . \ . \ .$$

and $\dfrac{3}{11}$ is
```
      .2727
 11)3.0000
   2 2
     80
     77
     30
     22
     80
     77
      3
```
or $\dfrac{3}{11} = 0.272727 \ . \ . \ .$

Notice the remainder 3 is equal to the original dividend. This indicates that the decimal quotient repeats itself.

We may use a shorthand notation to show that the decimal repeats.

$$\frac{1}{3} = 0.\overline{3} \qquad\qquad \frac{2}{3} = 0.\overline{6}$$

The bar means that the digits under the bar repeat endlessly.

$$\frac{3}{11} = 0.\overline{27} \quad \text{means } 0.272727\ldots$$

$$3\frac{1}{27} = 3.\overline{037} \quad \text{means } 3.037037037\ldots$$

Write $\frac{41}{33}$ as a repeating decimal using the "bar" notation. Check your answer in **31**.

MEASURING CONCENTRATIONS

The daily newspaper uses a very useful sort of measurement unit when it reports that the concentration of sulfur dioxide is 0.2 ppm in Detroit air during a smog alert. If we need to specify the concentration of very small relative amounts of material (as in air or water pollution or medicine dosages), we can write it as the ratio of the amount of substance added to the total amount of material. For example, adding 1 pound of salt to 1000 pounds of pure water will produce a concentration of

$$\frac{1 \text{ pound of salt}}{1000 \text{ pounds of water}} = \frac{1000 \times 1 \text{ pound of salt}}{1000 \times 1000 \text{ pounds of water}}$$

$$= \frac{1000 \text{ pounds of salt}}{1,000,000 \text{ pounds of water}}$$

or 1000 parts per million (ppm) of salt. Sea water contains about 35,000 ppm of dissolved solids. The allowable level of DDT in cow's milk is 0.05 ppm or

$$\frac{0.05 \text{ lb of DDT}}{1,000,000 \text{ lb of milk}} \quad \text{or} \quad \frac{0.05 \text{ ounces of DDT}}{1,000,000 \text{ ounces of milk}}$$

$$\text{or } \frac{0.05 \text{ grams of DDT}}{1,000,000 \text{ grams of milk}} \text{ (and so on)}$$

The actual amount of DDT in a cup or a quart or even a gallon of milk is measured in millionths of an ounce, but it is stored in human fat tissue and even amounts as tiny as 10 ppm can cause serious disorders.

31

$$33 \overline{) \begin{array}{r} 1.24 \\ 41.000 \end{array}}$$

$$\begin{array}{r} 33 \\ \hline 8\ 0 \\ 6\ 6 \\ \hline 1\ 40 \\ 1\ 32 \\ \hline 8 \end{array}$$

These remainders are the same and therefore further division will produce a repeat of the digits 24 in the quotient.

$$\frac{41}{33} = 1.242424 \ldots = 1.\overline{24}$$

Converting decimal numbers to fractions is fairly easy.

$$0.4 = \frac{4}{10} \text{ or } \frac{2}{5}$$

$$0.13 = \frac{13}{100}$$

$$0.275 = \frac{275}{1000} = \frac{13}{40} \text{ (reduced to lowest terms)}$$

$$0.035 = \frac{35}{1000} = \frac{7}{200} \text{ (reduced to lowest terms)}$$

Follow this procedure to convert decimal numbers to fractions:

Example

■Step 1: Write the digits to the right of the decimal point as the numerator in the fraction.

$$0.00325 = ?$$

$$\frac{325}{?}$$

■Step 2: In the denominator write 1 followed by as many zeros as there are decimal digits in the decimal number.

$$0.00325 = \frac{325}{100000}$$

5 digits   5 zeros

■Step 3: Reduce the fraction to lowest terms.

$$\frac{325}{100000} = \frac{13 \times 25}{4000 \times 25}$$

$$= \frac{13}{4000}$$

Write 0.036 as a fraction in lowest terms. Check your work in **32.**

DECIMAL–FRACTION EQUIVALENTS

Some fractions are used so often that it is worthwhile to list their decimal equivalents. Here are those used most often. Both rounded form and repeating decimal form are given. The lines marked ▲ are especially useful and should be memorized.

▲$\dfrac{1}{2}$ = 0.50

▲$\dfrac{1}{3}$ = 0.$\overline{3}$ or 0.33 rounded $\dfrac{2}{3}$ = 0.$\overline{6}$ or 0.67 rounded

▲$\dfrac{1}{4}$ = 0.25 $\dfrac{2}{4} = \dfrac{1}{2}$ = 0.50 $\dfrac{3}{4}$ = 0.75

▲$\dfrac{1}{5}$ = 0.20 $\dfrac{2}{5}$ = 0.40 $\dfrac{3}{5}$ = 0.60 $\dfrac{4}{5}$ = 0.80

$\dfrac{1}{6}$ = 0.1$\overline{6}$ or 0.17 rounded $\dfrac{5}{6}$ = 0.8$\overline{3}$ or 0.83 rounded

$\dfrac{1}{8}$ = 0.125 $\dfrac{3}{8}$ = 0.375 $\dfrac{5}{8}$ = 0.625 $\dfrac{7}{8}$ = 0.875

$\dfrac{1}{12}$ = 0.083$\overline{3}$ $\dfrac{5}{12}$ = 0.416$\overline{6}$ $\dfrac{7}{12}$ = 0.583$\overline{3}$ $\dfrac{11}{12}$ = 0.916$\overline{6}$

$\dfrac{1}{16}$ = 0.0625 $\dfrac{3}{16}$ = 0.1875 $\dfrac{5}{16}$ = 0.3121

$\dfrac{7}{16}$ = 0.4375 $\dfrac{9}{16}$ = .5625 $\dfrac{11}{16}$ = 0.6875

$\dfrac{13}{16}$ = 0.8125

$\dfrac{1}{20}$ = 0.05 $\dfrac{1}{25}$ = 0.04 $\dfrac{1}{50}$ = 0.02

32

$$0.036 = \underbrace{}_{\text{3 digits}} \frac{36}{\underbrace{1000}_{\text{3 zeros}}} = \frac{9}{250} \quad \text{(reduced to lowest terms)}$$

If the decimal number has a whole number portion, convert the decimal part to a fraction first and then add the whole number part. For example,

$$3.85 = 3 + 0.85$$

$$0.85 = \frac{85}{100} = \frac{17}{20} \quad \text{(reduced to lowest terms)}$$

$$3.85 = 3 + \frac{17}{20} = 3\frac{17}{20}$$

Write the following decimal numbers as fractions in lowest terms.

(a) 0.0075   (b) 2.08   (c) 3.11

Check your work in **33**.

HOW TO WRITE A REPEATING DECIMAL AS A FRACTION

A repeating decimal is one in which some sequence of digits is endlessly repeated. For example, $0.333 \ldots = 0.\overline{3}$ and $0.272727 \ldots = 0.\overline{27}$ are repeating decimals. The bar over the number is a shorthand way of showing that those digits are repeated.

What fraction is equal to $0.\overline{3}$? To answer this form a fraction with numerator equal to the repeating digits and denominator equal to a number formed with the same number of 9s.

$$0.\overline{3} = \frac{3}{9} = \frac{1}{3}$$

$$0.\overline{27} = \frac{27}{99} = \frac{3}{11} \qquad \text{Two digits in } 0.\overline{27}, \text{ therefore use 99 as the denominator.}$$

$$0.\overline{123} = \frac{123}{999} = \frac{41}{333} \qquad \text{Three digits in } 0.\overline{123}, \text{ therefore use 999 as the denominator.}$$

33

(a) $0.0075 = \dfrac{75}{10000} = \dfrac{3}{400}$

(b) $2.08 = 2 + \dfrac{8}{100} = 2 + \dfrac{2}{25} = 2\dfrac{2}{25}$

(c) $3.11 = 3 + \dfrac{11}{100} = 3\dfrac{11}{100}$

Now turn to **34** for a set of practice problems on decimal fractions.

34

Problem Set 3: Decimal Fractions

A. Write as decimal numbers (round to two decimal digits):

$\dfrac{1}{2}$ = _____ $\dfrac{1}{3}$ = _____ $\dfrac{2}{3}$ = _____

$\dfrac{1}{4}$ = _____ $\dfrac{2}{4}$ = _____ $\dfrac{3}{4}$ = _____

$\dfrac{1}{5}$ = _____ $\dfrac{2}{5}$ = _____ $\dfrac{3}{5}$ = _____

$\dfrac{4}{5}$ = _____ $\dfrac{1}{6}$ = _____ $\dfrac{5}{6}$ = _____

$\dfrac{1}{7}$ = _____ $\dfrac{2}{7}$ = _____ $\dfrac{3}{7}$ = _____

$\dfrac{1}{8}$ = _____ $\dfrac{3}{8}$ = _____ $\dfrac{5}{8}$ = _____

$\dfrac{7}{8}$ = _____ $\dfrac{1}{10}$ = _____ $\dfrac{2}{10}$ = _____

$\dfrac{3}{10}$ = _____ $\dfrac{1}{12}$ = _____ $\dfrac{2}{12}$ = _____

$\dfrac{3}{12}$ = _____ $\dfrac{5}{12}$ = _____ $\dfrac{7}{12}$ = _____

$\dfrac{11}{12}$ = _____ $\dfrac{1}{16}$ = _____ $\dfrac{3}{16}$ = _____

$\dfrac{5}{16}$ = _____ $\dfrac{7}{16}$ = _____ $\dfrac{9}{16}$ = _____

$\dfrac{11}{16}$ = _____ $\dfrac{13}{16}$ = _____ $\dfrac{15}{16}$ = _____

B. Write as a fraction in lowest terms:

0.3 = _____   0.75 = _____   0.44 = _____

0.8 = _____   0.6 = _____   0.025 = _____

0.4 = _____   1.3 = _____   2.25 = _____

2.05 = _____   3.16 = _____   1.125 = _____

3.22 = _____   2.04 = _____   0.075 = _____

10.875 = _____   0.0007 = _____   0.0012 = _____

0.34 = _____   11.0105 = _____   6.0020 = _____

4.115 = _____   0.35 = _____   0.955 = _____

C. Solve:

1. What decimal fraction of .5 is .6?

2. Find 3.4 of 120.

3. If .07 of some number is .315, find the number.

4. What decimal part of 2.5 is 42.5?

5. Find a number such that .78 of it is .390.

D. Brain Boosters

1. $2\frac{3}{5} + 1.785 =$ _____ $\frac{1}{5} + 1.57 =$ _____

$3\frac{7}{8} - 2.4 =$ _____ $1\frac{3}{16} - 0.4194 =$ _____

$2\frac{1}{2} \times 3.15 =$ _____ $1\frac{3}{25} \times 2.05 =$ _____

$3\frac{4}{5} \div 2.65 =$ _____ $1\frac{5}{16} \div 4.2 =$ _____

2. On four examinations in his history course, Denny scored 73.7, 81.4, 63.6, and 75.6. Find his average exam grade. (Add the scores and divide by 4, the number of test scores.)

3. Verify by dividing that $\frac{1}{7} = 0.\overline{142857}$, a repeating decimal. Express $\frac{2}{7}, \frac{3}{7}, \frac{4}{7}, \frac{5}{7},$ and $\frac{6}{7}$ as repeating decimals. What do you notice about these decimal numbers?

4. If seven avocados cost $3, what would be the selling price of one?

5. If Andy is paid $2.25 per hour, how much money will he receive for $3\frac{1}{2}$ hours of work?

6. One tablet of calcium pantothenate contains 0.5 grams. How much is contained in $2\frac{3}{4}$ tablets? How many tablets are needed to make up 2.6 grams?

The answers to these problems are on page 231. When you have had the practice you need, turn to **35** for a self-test.

35

Chapter 3 Self-Test

1. $6.2 + 13.045 =$ _____

2. $41.3 + 9.86 =$ _____

3. $16 + 3.407 + 21.744 =$ _____

4. $76 - 7.93 =$ _____

5. $4.27 - 3.8 =$ _____

6. $237.4 - 65.87 =$ _____

7. $90 - 14.85 =$ _____

8. $8.1 \times 2.04 =$ _____

9. $5.6 \times 30 =$ _____

10. $30.4 \times 1.005 =$ _____

11. $8 \div 4.2$ (round to two decimal digits) = _____

12. $14.2 \div 0.075$ (round to two decimal digits) = _____

13. $83.07 \div 104.6$ (round to three decimal digits) = _____

14. Write 0.56 as a fraction in lowest terms. _____

15. Write 3.248 as a fraction in lowest terms. _____

16. Write 32.13 as a fraction in lowest terms. _____

17. Write $\dfrac{7}{16}$ as a decimal. _____

18. Write $3\dfrac{5}{8}$ as a decimal. _____

19. What part of 3.8 is 4.56? _____

20. What part of 7.0 is 4.2? _____

21. Find 0.25 of 4.8. _____

22. Find 0.65 of 23. _____

23. Find 2.45 of 3.5. _____

24. Find a number such that 0.35 of it is 2.45. _____

25. Find a number such that 1.4 of it is 17.5. _____

The answers to these problems are on page 235.

CHAPTER FOUR

Percent

Preview

Objectives	Where to Go for Help	
Upon successful completion of this chapter you will be able to:	Page	Frame

1. Write fractions and decimal fractions as percents.

    (a) Write $1\frac{7}{8}$ as a percent. _____   177   **1**

    (b) Write 0.45 as a percent. _____   177   **1**

2. Convert percents to decimal and fraction form.

    (a) Write $37\frac{1}{2}\%$ as a decimal. _____   177   **1**

    (b) Write 44% as a fraction. _____   177   **1**

3. Solve problems involving percent.

    (a) Find 35% of 16. _____   187   **12**

    (b) Find 120% of 45. _____   187   **12**

    (c) What percent of 18 is 3? _____   187   **12**

    (d) What percent of $1\frac{2}{3}$ is $\frac{1}{2}$? _____   187   **12**

    (e) What percent of 0.6 is 0.25? _____   187   **12**

    (f) 70% of what number is 56? _____   187   **12**

4. Solve practical problems involving percent.

		Page	Frame

(a) How much money does a salesman earn on a $240 sale if his commission is 15%? _____   203   **24**

(b) A camera normally selling for $149.50 is on sale at a discount of 25%. What is its sale price? _____   203   **24**

(c) After a 10% discount a book sells for $5.40. What was the original price? _____   203   **24**

(d) A coat sells for $39.75 plus 6% sales tax. What is the total cost? _____   203   **24**

(e) What is the interest paid on a $1000 bank loan at $6\frac{1}{2}\%$ for 24 months? _____   203   **24**

If you are certain you can work all of these problems correctly, turn to page 215 for a self-test. If you want help with any of these objectives or if you cannot work one of the preview problems, turn to the page indicated. Super-students who are eager to learn everything in this unit will turn to frame **1** and begin work there.

ANSWERS TO PREVIEW PROBLEMS

1. (a) 187.5%
    (b) 45%

2. (a) 0.375
    (b) $\frac{11}{25}$

3. (a) 5.6
    (b) 54
    (c) $16\frac{2}{3}\%$
    (d) 30%
    (e) $41\frac{2}{3}\%$
    (f) 80

4. (a) $36.00
    (b) $112.12
    (c) 5.94
    (d) $42.14
    (e) $130

Percent

1

Sitting at that desk, doing math in the abstract, make-believe world of school, Sally finds it easy to lose her sense of the way it ought to be. For Sally mathematics is a guessing game played only in school, with correct guesses rewarded by a pat on the head and wrong guesses followed by another guess. When she does math problems for real (while window shopping, buying something, or counting change) she stops the wild guessing and tries to reason them out. When she gets a bit older and worries about buying a car or a house, getting a loan, paying taxes, buying on an installment plan, earning interest on savings, or shopping for bargains, she'll find that in order to reason out these new problems she must understand the concept of percent. Wild, random guesses then will mean she's going to lose a lot of money.

The word <u>percent</u> comes from the Latin words <u>per</u> <u>centum</u> meaning "by the hundred" or "for every hundred." A number expressed as a percent is being compared with a second number called the standard or <u>base</u> by dividing the base into 100 equal parts and writing the comparison number as so many hundredths of the base.

For example, what part of the base or standard length is length A?

base	A

We could answer the question with a fraction or a decimal or a percent. First, divide the base into 100 equal parts. Then compare A with it. The length of A is 40 parts out of the 100 parts that make up the base.

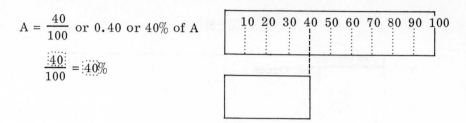

$$A = \frac{40}{100} \text{ or } 0.40 \text{ or } 40\% \text{ of A}$$

$$\frac{40}{100} = 40\%$$

Thus 40% means 40 parts in 100 or $\frac{40}{100}$.

What part of this base is length B? Answer with a percent.

B

Turn to **2** to check your answer.

2

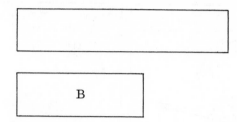

B is $\frac{60}{100}$ or 60%.

B

Of course the compared number may be larger than the base.

base

C

In this case, divide the base into 100 parts and extend it in length.

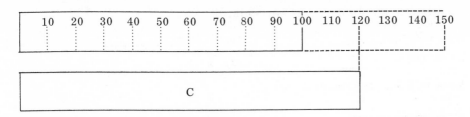

The length of C is 120 parts out of the 100 parts that make up the base.

$$\text{C is } \frac{120}{100} \text{ or } 120\% \text{ of A.}$$

Because both our number system and our money system are based on ten and multiples of ten, it is very handy to write comparisons in hundredths or percent.

What part of $1.00 is 50 cents? Write your answer as a fraction, as a decimal, and as a percent. Check in **3** for the answer.

3

50¢ is what part of $1.00?

$$\frac{50\cancel{c}}{100\cancel{c}} = \frac{50}{100} = .50 = 50\% \qquad 50¢ \text{ is equal to } 50\% \text{ of } \$1.00.$$

We may also write $50¢ = \frac{1}{2}$ of $1.00 or 50¢ = 0.50 of $1.00. Fractions, decimals, and percents are alternative ways to talk about a comparison of two numbers.

What percent of 10 is 2? Write 2 as a fraction of 10, rename it as a fraction with denominator equal to 100, then write it as a percent. When you have completed this, go to **4**.

4

$$\frac{2}{10} = \frac{2 \times 10}{10 \times 10} = \frac{20}{100} = 20\%$$

How do you rewrite a decimal number as a percent? The procedure is simply to multiply the decimal number by 100%.

$$0.60 = 0.60 \times 100\% = 60\% \qquad \begin{array}{r} .60 \\ \times\ 100 \\ \hline 60.00 \end{array} = 60\%$$

$$0.375 = 0.375 \times 100\% = 37.5\%$$

$$3.4 = 3.4 \times 100\% = 340\%$$

$$0.02 = 0.2 \times 100\% = 2\%$$

Rewrite the following as percents.

(a) 0.70 = _____   (b) 1.25 = _____

(c) 0.001 = _____   (d) 3 = _____

Look in **5** for the answers.

Does "of" always mean multiply?

In ordinary language it can mean many things, but in percent problems it almost always means that a multiplication is coming up.

5

(a) $0.70 = 0.70 \times 100\% = 70\%$

(b) $1.25 = 1.25 \times 100\% = 125\%$

(c) $0.001 = 0.001 \times 100\% = 0.1\%$

(d) $3 = 3 \times 100\% = 300\%$

Notice in each of these that multiplication by 100% has the effect of moving the decimal point two digits to the right.

$$0.70 \text{ becomes } 70.\% \text{ or } 70\%$$

$$1.25 \text{ becomes } 125.\% \text{ or } 125\%$$

$$0.001 \text{ becomes } 0.1\%$$

$$3 = 3.00 \text{ becomes } 300.\% \text{ or } 300\%$$

To rewrite a fraction as a percent, we can always rename it as a fraction with 100 as the denominator.

$$\frac{3}{20} = \frac{3 \times 5}{20 \times 5} = \frac{15}{100} = 15\%$$

However, the easiest way is to convert the fraction to decimal form by dividing the denominator into the numerator and then multiplying by 100%.

$$\frac{1}{2} = 0.50 = 0.50 \times 100\% = 50\%$$

$$2\overline{)1.00} \quad .50$$

$$\frac{3}{4} = 0.75 = 0.75 \times 100\% = 75\%$$

$$4\overline{)3.00} \quad .75$$

$$\frac{3}{20} = 0.15 = 0.15 \times 100\% = 15\%$$

$$\begin{array}{r} .15 \\ 20\overline{)3.00} \\ \underline{2\ 0} \\ 1\ 00 \\ \underline{1\ 00} \end{array}$$

$$1\frac{7}{20} = \frac{27}{20} = 1.35 = 1.35 \times 100\% = 135\%$$

$$\begin{array}{r} 1.35 \\ 20\overline{)27.00} \\ \underline{20} \\ 7\ 0 \\ \underline{6\ 0} \\ 1\ 00 \\ \underline{1\ 00} \end{array}$$

Rewrite $\frac{5}{16}$ as a percent. Check your answer in **6**.

6

$$\frac{5}{16} = 0.3125 = 0.3125 \times 100\% = 31.25\%$$

This is often written as $31\frac{1}{4}\%$.

$$\begin{array}{r} .3125 \\ 16\overline{)5.0000} \\ \underline{4\ 8} \\ 20 \\ \underline{16} \\ 40 \\ \underline{32} \\ 80 \\ \underline{80} \end{array}$$

Some fractions cannot be converted to exact decimals. For example,

$$\frac{1}{3} = 0.333\ .\ .\ .$$

where the 3s continue to repeat endlessly. We can round to get an approximate percent:

$$\frac{1}{3} = 0.333 \times 100\% = 33.3\%$$

Or we can convert to a fraction with 100 as denominator:

$$\frac{1}{3} = \frac{?}{100} \qquad \frac{1}{3} = \frac{1 \times 33\frac{1}{3}}{3 \times 33\frac{1}{3}} = \frac{33\frac{1}{3}}{100} = 33\frac{1}{3}\%$$

Rewrite $\frac{1}{6}$ as a percent. The answer is in **8**.

7

(a) $\dfrac{7}{5} = 1.4$ $\qquad$ $1.4 \times 100\% = 140\%$

(b) $\dfrac{2}{3} = \dfrac{?}{100}$ $\qquad$ $\dfrac{2}{3} = \dfrac{2 \times 33\frac{1}{3}}{3 \times 33\frac{1}{3}} = \dfrac{66\frac{2}{3}}{100} = 66\frac{2}{3}\%$

(c) $3\dfrac{1}{8} = \dfrac{25}{8} = 3.125$ $\qquad$ $3.125 \times 100\% = 312.5\%$

(d) $\dfrac{5}{12} = \dfrac{?}{100} = \dfrac{5 \times 8\frac{1}{3}}{12 \times 8\frac{1}{3}} = \dfrac{40\frac{5}{3}}{100} = \dfrac{41\frac{2}{3}}{100} = 41\frac{2}{3}\%$

In order to use percent in solving practical problems it is often necessary to change a percent to a decimal number. The procedure is to divide by 100%.

$$50\% = \dfrac{50\%}{100\%} = \dfrac{50}{100} = 0.50 \qquad\qquad 100\overline{)50.0}^{\,.5}$$

$$5\% = \dfrac{5\%}{100\%} = \dfrac{5}{100} = 0.05 \qquad\qquad 100\overline{)5.00}^{\,.05}$$

$$0.2\% = \dfrac{0.2\%}{100\%} = \dfrac{0.2}{100} = 0.002 \qquad\qquad 100\overline{)0.200}^{\,.002}$$

Fractions may be part of the percent number. If so, it is easiest to reduce to a decimal number and round if necessary.

$$6\tfrac{1}{2}\% = \dfrac{6\frac{1}{2}\%}{100\%} = \dfrac{6.5}{100} = 0.065 \qquad\qquad 100\overline{)6.500}^{\,.065}$$

$$33\tfrac{1}{3}\% = \dfrac{33\frac{1}{3}\%}{100\%} = \dfrac{33\frac{1}{3}}{100} = \dfrac{33.3}{100} = 0.333 \text{ (rounded)}$$

Now you try a few. Write these as decimal numbers.

(a) $4\% = $ _____ $\qquad\qquad$ (b) $0.5\% = $ _____

(c) $16\dfrac{2}{3}\% = $ _____ $\qquad\qquad$ (d) $79\dfrac{1}{4}\% = $ _____

Our answers are in **9**.

10

(a) $72\% = \dfrac{72\%}{100\%} = \dfrac{72}{100} = \dfrac{18 \times 4}{25 \times 4} = \dfrac{18}{25}$

(b) $16\dfrac{1}{2}\% = \dfrac{16\frac{1}{2}\%}{100\%} = \dfrac{16\frac{1}{2}}{100} = \dfrac{\frac{33}{2}}{100} = \dfrac{33}{200}$

(c) $240\% = \dfrac{240\%}{100\%} = \dfrac{240}{100} = \dfrac{12 \times 20}{5 \times 20} = \dfrac{12}{5} = 2\dfrac{2}{5}$

(d) $7\dfrac{1}{2}\% = \dfrac{7\frac{1}{2}\%}{100\%} = \dfrac{7\frac{1}{2}}{100} = \dfrac{\frac{15}{2}}{100} = \dfrac{15}{200} = \dfrac{3}{40}$

Now turn to **11** for a set of practice problems on what you have learned in this chapter so far.

11

Problem Set 1: Numbers and Percent

A. Write each number as a percent:

$0.40 = $ _____ $0.10 = $ _____ $0.95 = $ _____

$0.03 = $ _____ $0.3 = $ _____ $0.015 = $ _____

$0.6 = $ _____ $0.01 = $ _____ $1.2 = $ _____

$4.56 = $ _____ $2.25 = $ _____ $7.75 = $ _____

$0.003 = $ _____ $3.0 = $ _____ $0.8 = $ _____

$5.5 = $ _____ $4 = $ _____ $6.04 = $ _____

$10 = $ _____ $0.335 = $ _____

B. Write each fraction as a percent:

$\dfrac{1}{5} = $ _____ $\dfrac{3}{4} = $ _____ $\dfrac{7}{10} = $ _____

$\dfrac{7}{20} = $ _____ $\dfrac{3}{2} = $ _____ $\dfrac{1}{4} = $ _____

$\dfrac{1}{10} = $ _____ $\dfrac{1}{2} = $ _____ $\dfrac{3}{8} = $ _____

$\dfrac{3}{5} = $ _____ $\dfrac{7}{4} = $ _____ $\dfrac{11}{5} = $ _____

$1\dfrac{4}{5} =$ _____ $\dfrac{9}{10} =$ _____ $\dfrac{1}{3} =$ _____

$2\dfrac{1}{6} =$ _____ $\dfrac{2}{3} =$ _____ $\dfrac{11}{16} =$ _____

$\dfrac{23}{12} =$ _____ $3\dfrac{3}{10} =$ _____

C. Write each percent as a decimal number:

$7\% =$ _____ $3\% =$ _____ $56\% =$ _____

$15\% =$ _____ $1\% =$ _____ $7\dfrac{1}{2}\% =$ _____

$90\% =$ _____ $200\% =$ _____ $0.3\% =$ _____

$0.07\% =$ _____ $.25\% =$ _____ $150\% =$ _____

$1\dfrac{1}{2}\% =$ _____ $6\dfrac{1}{3}\% =$ _____ $\dfrac{1}{2}\% =$ _____

$12\dfrac{1}{4}\% =$ _____ $125\dfrac{1}{2}\% =$ _____ $66\dfrac{2}{3}\% =$ _____

$30\dfrac{1}{2}\% =$ _____ $8\dfrac{1}{2}\% =$ _____

D. Write each percent as a fraction in lowest terms:

$10\% =$ _____ $65\% =$ _____ $50\% =$ _____

$20\% =$ _____ $25\% =$ _____ $8\% =$ _____

$90\% =$ _____ $135\% =$ _____ $3\% =$ _____

$12\% =$ _____ $\dfrac{1}{2}\% =$ _____ $0.03\% =$ _____

$4.5\% =$ _____ $220\% =$ _____ $1\dfrac{1}{2}\% =$ _____

$33\dfrac{1}{3}\% =$ _____ $7\dfrac{3}{4}\% =$ _____ $6\dfrac{1}{2}\% =$ _____

$16\dfrac{2}{3}\% =$ _____ $3\dfrac{1}{8}\% =$ _____

The answers to these problems are on page 232. When you have had the practice you need, either return to the preview for this chapter on page 175 or continue in frame **12** with the study of problems involving percent.

Percent Problems

12

In all of your work with percent you will find that there are three basic types of problems. These three form the basis for all percent problems that arise in business, industry, science, or other areas. All of these problems involve three quantities:

- the <u>base</u> or <u>total</u> amount or standard used for a comparison

- the <u>percentage</u> or part being compared with the base or total

- the <u>percent</u> or <u>rate</u> which indicates the relationship of the percentage to the base (the part to the total)

All three basic percent problems involve finding one of these three quantities when the other two are known. In every problem follow these three steps:

■ Step 1: Translate the problem sentence into a math statement. (If you have not already read the section on word problems in Chapter 2, do so now. It starts on page 120, frame **51.**) For example, the question "30% of what number is 16?" should be translated:

$$\text{30\% of what number is 16?}$$
$$30\% \times \qquad \square \qquad = 16$$

$$30\% \times \square = 16$$

In this case, 30% is the percent or rate; $\square$, the unknown quantity, is the total or base; and 16 is the percentage or part of the total. Notice that the words and phrases in the problem become math symbols. The word "of" is translated <u>multiply</u>. The word "is" (and similar verbs such as "will be" and "becomes") are translated <u>equals</u>. Use a $\square$ or letter of the alphabet or ? for the unknown quantity you are asked to find.

■ Step 2: It will be helpful if you <u>label</u> which numbers are the base or total (T), the percent (%), and the percentage or part (P).

$$30\% \times \square = 16$$
$$\quad \% \quad T \quad P$$

■ Step 3: <u>Rearrange</u> the equation so that the unknown quantity is alone on the left of the equal sign and the other quantities are on the right of the equal sign. The equation $30\% \times \square = 16$ becomes

$$\square = 16 \div 30\% = \frac{16}{30\%} = \frac{16}{.30}$$

The following Equation Finder may help you do this arranging. By dividing this circle we can write three equivalent equations, each describing one of the three basic percent problems.

$A = B \times C$

$B = \dfrac{A}{C}$

$\dfrac{A}{B} = C$

or $C = \dfrac{A}{B}$

The three equations $A = B \times C$, $B = \dfrac{A}{C}$, and $C = \dfrac{A}{B}$ are all equivalent.

■Step 4: Make a reasonable estimate of the answer. Guess, but guess carefully. Good guessing is an art.

■Step 5: Solve the problem by doing the arithmetic.

◊ BE CAREFUL. Never do arithmetic, never multiply or divide, with percent numbers. All percents must be rewritten as fractions or decimals before you can use them in a multiplication or division.

■Step 6: Check your answer against the original guess. Are they the same or at least close? If they do not agree, at least roughly, you have probably made a mistake and should repeat your work.

■ Step 7: <u>Double-check</u> by putting the answer number you have found back into the original problem or equation to see if it makes sense. If possible, use the answer to calculate one of the other numbers in the equation as a check.

Now let's look very carefully at each type of problem. We'll explain each, give examples, show you how to solve them, and work through a few together.   Turn to **13**.

13

Type <u>1 problems</u> are usually stated in the form "Find 30% of 50" or "What is 30% of 50?" or "30% of 50 is what number?"

■ Step 1: Translate. $30\% \times 50 = \square$

■ Step 2: Label. $\% \times T = P$

■ Step 3: Rearrange. $\square = 30\% \times 50$

You complete the calculation and find $\square$.

   $\square = \underline{\hspace{2cm}}$

Choose an answer:

(a)	150	Go to **14.**
(b)	15	Go to **15.**
(c)	1500	Go to **16.**

How can you have more than 100% of something? All of it is all of it, right?

Percent relates size of something to a standard. If it is larger than the standard, it is more than 100% of the standard.

14

You answered that 30% of 50 = 150 and this is not correct. Be certain you do the following in solving this problem:

1. Guess at the answer. A careful, educated guess or estimate is an excellent check on your work. Never work a problem until you know roughly the size of the answer. 30% is roughly $\frac{1}{3}$. What is $\frac{1}{3}$ of 50, approximately?

2. Never multiply by a percent number. Before you multiply 30% × 50 you must write 30% as a decimal number. (If you need help with this turn to **8**.)

Use these hints to solve the problem. Then return to **13** and choose a better answer.

15

Right you are!

■ Step 4: <u>Guess</u>. The next step is to make an educated guess at the answer. For example,

$$30\% \text{ of } 50 \text{ is roughly } \frac{1}{3} \text{ of } 50 \text{ or about } 17$$

Your answer will be closer to 17 than to 2 or 100.

■ Step 5: <u>Solve</u>. Never multiply by a percent number. Percent numbers are not arithmetic numbers. Before you multiply write the percent number as a decimal number.

$$30\% = \frac{30\%}{100\%} = \frac{30}{100} = 0.30$$
$$30\% \times 50 = 0.30 \times 50 = 15$$

■ Step 6: <u>Check</u>. The guess (17) and the answer (15) are not exactly the same but they are reasonably close. The answer seems acceptable.

Don't be intimidated by numbers. If the problem involves very large or very complex numbers reduce it to a simpler problem. The problem

$$\text{"Find } 14\frac{7}{32}\% \text{ of } 6.4\text{"}$$

may look difficult until you realize that it is essentially the same problem as "Find 10% of 6" which is fairly easy.

Before you begin any actual arithmetic problem involving percent you should:

- know roughly the size of the answer,

- have a plan for solving the problem based on a simpler problem,

- always change percents to decimals or fractions before multiplying or dividing with them.

Now try this problem:

Find $8\frac{1}{2}\%$ of 160.

Check your answer in **17**.

16

Your answer is incorrect. First, it is important that you make an educated guess at the answer. Never work a problem before you know roughly the size of the answer. In this case, 30% of 50 is roughly $\frac{1}{3}$ of 50.

𝕭𝖊𝖜𝖆𝖗𝖊 ⬦Second, <u>never</u> multiply by a percent number. Before you multiply 30% × 50 you must write 30% as a decimal number. If you need help in writing a percent as a decimal number turn to frame **8**. Otherwise, return to **13** and try again.

17

- **Step 1: Translate.** $8\frac{1}{2}\% \times 160 = \square$

- **Step 2: Label.** $\% \times T = P$

- **Step 3: Rearrange.** $\square = 8\frac{1}{2}\% \times 160$

- **Step 4: Guess.** $8\frac{1}{2}\%$ is close to 10%. 10% of 160 is $\frac{1}{10}$ of 160 or 16.

- **Step 5: Solve.** $8\frac{1}{2}\% = \dfrac{8\frac{1}{2}\%}{100\%} = \dfrac{8.5}{100} = 0.085$

                              $\square = 0.085 \times 160 = 13.6$

- **Step 6: Check.**   13.6 is approximately equal to 16, our original estimate.

Now try these for practice.

(a) Find 2% of 140. _____

(b) 35% of 20 = _____

(c) $7\frac{1}{4}$% of $1000 = _____

(d) What is $5\frac{1}{3}$% of 3.3? _____

(e) 120% of 15 is what number? _____

The step-by-step answers are in **18.**

What does "%" mean? Where did that goofy-looking symbol come from?

It means "100." The word "percent" comes from the Latin words meaning "for every hundred." It started as 100, then 1oo, then $\frac{o}{o}$ and finally % or %.

18

(a) 2% of 140 = ?

  2% × 140 = ☐

  ☐ = 2% × 140

  ☐ = .02 × 140

  ☐ = 2.8

Guess: 10% or $\frac{1}{10}$ of 140 = 14.

  2% of 140 would be about 3.

  $(2\% = \dfrac{2\%}{100\%} = \dfrac{2}{100} = 0.02)$

Check: 2.8 is roughly equal to 3, our original guess.

(b) 35% of 20 = ?

  35% × 20 = ☐

  ☐ = 0.35 × 20

  ☐ = 7

Guess: 35% is roughly $\frac{1}{3}$ and $\frac{1}{3}$ of 20 is about 6 or 7.

Check: Our guess (6 or 7) is very close to the answer (7).

(c) $7\frac{1}{4}\%$ of $1000 = ?$

$$7\frac{1}{4}\% \times \$1000 = \square$$

$$\square = 7\frac{1}{4}\% \times \$1000$$

$$\square = 0.0725 \times \$1000$$

$$\square = \$72.50$$

Guess: 10%, or $\frac{1}{10}$, of $1000 is $100. The answer should be less than $100.

$$7\frac{1}{4}\% = 7.25\% = \frac{7.25}{100} = 0.0725$$

$$\begin{array}{r} .0725 \\ \times\ 1000 \\ \hline 72.5000 \end{array}$$

Check: The answer, $72.50, is a bit less than the guess of about $100.

(d) $5\frac{1}{3}\%$ of $3.3 = ?$

$$5\frac{1}{3}\% \times 3.3 = \square$$

$$\square = 5\frac{1}{3}\% \times 3.3$$

$$\square = \frac{16}{300} \times 3.3$$

$$\square = \frac{16}{300} \times \frac{33}{10} = \frac{528}{3000}$$

$$\square = 0.176$$

Guess: 10%, or $\frac{1}{10}$, of 3 is 0.3. 5% of 3 would be half of this or 0.15. A good guess at the answer would be 0.15.

$$5\frac{1}{3}\% = \frac{5\frac{1}{3}}{100} = \frac{\frac{16}{3}}{100} = \frac{16}{300}$$

Check: The answer, 0.176, is reasonably close to the guess, 0.15.

(e) 120% of 15 = ?

$$120\% \times 15 = \square$$

$$\square = 120\% \times 15$$

$$\square = 1.2 \times 15$$

$$\square = 18$$

Guess: 100% of 15 is 15 so the answer is certainly more than 15. 200% of 15 is twice 15 or 30. A good guess would be that the answer is between 15 and 20.

$$120\% = \frac{120}{100} = 1.2$$

Check: The answer, 18, is between 15 and 20, just as we expected our answer to be.

Type 2 problems require that you find the rate or percent. Problems of this kind are usually stated "7 is what percent of 16?" or "Find what percent 7 is of 16" or "What percent of 16 is 7?"

■Step 1: Translate.   What percent of 16 is 7?

$$\square\% \times 16 = 7$$

■Step 2: Label. $\% \times T = P$

All of the problem statements are equivalent to this equation.

■Step 3: Rearrange. $\square\% = \dfrac{7}{16}$

To rearrange the equation and solve for $\%$ notice that it is in
the form

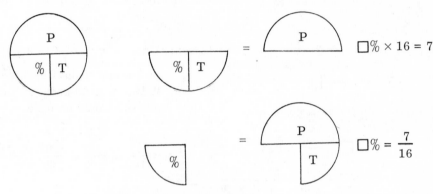

Therefore, $\square\% = \dfrac{7}{16}$. 16 is the total amount or base and 7 is
the part of the base being described.

Solve this last equation. Check your answer in **19.**

19

Guess: $\dfrac{7}{16}$ is very close to $\dfrac{8}{16}$ or $\dfrac{1}{2}$ or 50%. The answer will be a
little less than 50%.

$$\square\% = \frac{7}{16} = \frac{7}{16} \times 100\% = \frac{700}{16}\%$$

$$\square\% = 43\frac{3}{4}\%$$

$$\begin{array}{r} 43 = 43\frac{12}{16} = 43\frac{3}{4} \\ 16\overline{)700} \\ \underline{64} \\ 60 \\ \underline{48} \\ 12 \end{array}$$

Check: The answer, $43\frac{3}{4}\%$, is reasonably close to our preliminary
guess of about 50%.

If you had trouble converting $\dfrac{7}{16}$ to a percent, you should review this
process by turning back to frame **5.**

The solution to a Type 2 problem will be a fraction or decimal number that must be converted to a percent.

Try these problems for practice.

(a) What percent of 40 is 16?

(b) Find what percent 65 is of 25.

(c) $6.50 is what percent of $18.00?

(d) What percent of 2 is 3.5?

(e) $10\frac{2}{5}$ is what percent of 2.6?

Check your work in **20**.

HOW TO MISUSE PERCENT

1. In general you cannot add, subtract, multiply, or divide percent numbers. Percent helps you compare two numbers. It cannot be used in the normal arithmetic operations.

   For example, if 60% of class 1 earned A grades and 50% of class 2 earned A grades, what was the total percent of A grades for the two classes? The answer is that you cannot tell unless you know the number of students in each class.

2. In advertisements designed to trap the unwary, you might hear that "children had 23% fewer cavities when they used . . ." or "50% more doctors smoke . . . ."

   <u>Fewer</u> than what? Fewer than the worst dental health group the advertiser could find? Fewer than the national average?

   <u>More</u> than what? More than a year ago? More than nurses? More than infants?

   There must be some reference or base given for the percent number to have any meaning at all.

BEWARE of people who misuse percent!

20

(a) $\square\% \times 40 = 16$ Guess: $\dfrac{16}{40}$ is about $\dfrac{1}{3}$ or roughly 33%.

$\qquad \square\% = \dfrac{16}{40}$ $\% = \dfrac{P}{T}$

$\qquad \dfrac{16}{40} = \dfrac{16}{40} \times 100\% = \dfrac{1600}{40}\%$

$\qquad \square\% = 40\%$ Check: 40% is reasonably close to 33%.

Double-check: $40\% \times 40 = ?$

$\qquad .40 \times 40 = 16$

(b) $\square\% \times 25 = 65$ Guess: $\dfrac{65}{25}$ is more than 2 and 2 is

$\qquad \square\% = \dfrac{65}{25}$ 200%. The answer will be over 200%.

$\qquad \dfrac{65}{25} = \dfrac{65}{25} \times 100\% = 260\%$

$\qquad \square\% = 260\%$ Check: The answer agrees with the guess.

Double-check: $260\% \times 25 = ?$

$\qquad 2.60 \times 25 = 65.00 = 65$

The most difficult part of this problem is deciding whether the percent needed is found from $\dfrac{65}{25}$ or $\dfrac{25}{65}$. There is no magic to it. If you read the problem very carefully you will see that it speaks of 65 as a part "of 25." The base or total is 25. The percentage or part is 65.

(c) $\$6.50 = \square\% \times \$18.00$

or $\square\% \times \$18.00 = \$6.50$ $\% = \dfrac{P}{T}$

$\qquad \square\% = \dfrac{\$6.50}{\$18.00}$ Guess: $\$6.50$ is about $\dfrac{1}{3}$ of $\$18.00$

and $\dfrac{1}{3}$ is roughly 30%.

$\qquad \square\% = \dfrac{6.50}{18.00} = \dfrac{6.50}{18.00} \times 100\%$

$\qquad \square\% = \dfrac{650}{18}\% = 36\dfrac{2}{18}\%$

$\qquad \square\% = 36\dfrac{1}{9}\%$ Check: $36\dfrac{1}{9}\%$ is reasonably close to the guess of 30%.

(c) continued

Double-check: $36\frac{1}{9}\% \times \$18 = ?$

$$\frac{325}{900} \times 18 = 6.5$$

(d) $\square\% \times 2 = 3.5$

$\% = \dfrac{P}{T}$

$\square\% = \dfrac{3.5}{2}$

Guess: $\dfrac{3}{2}$ is $1\dfrac{1}{2}$ or 1.5 and 1.5 is 150%. The answer will be something more than 150%.

$\square\% = \dfrac{3.5}{2} \times 100\% = \dfrac{350}{2}\%$

$\square\% = 175\%$

Check: The answer and the guess are roughly the same.

Double-check: $175\% \times 2 = ?$

$$1.75 \times 2 = 3.50$$

(e) $10\dfrac{2}{5} = \square\% \times 2.6$

$\% = \dfrac{P}{T}$

or $\square\% \times 2.6 = 10\dfrac{2}{5}$

Guess: $\dfrac{10}{2}$ is 5 and 5 is 500%.

$\square\% = \dfrac{10\dfrac{2}{5}}{2.6}$

$\square\% = \dfrac{10.4}{2.6} = \dfrac{10.4}{2.6} \times 100\% = \dfrac{1040}{2.6}\%$

$\square\% = 400\%$

Check: The answer and the guess are roughly the same.

Double check: $400\% \times 2.6 = ?$

$$4 \times 2.6 = 10.4 = 10\dfrac{2}{5}$$

Type 3 problems require that you find the total given the percent and the percentage or part. These problems are usually stated like this: "8.7 is 30% of what number?" or "Find a number such that 30% of it is 8.7" or "8.7 is 30% of a number; find the number" or "30% of what number is equal to 8.7?"

■ Step 1: Translate.   30% of what number is equal to 8.7?

$$30\% \times \square = 8.7$$

■ Step 2: Label. $\% \times T = P$

■Step 3: Rearrange. $\square = \dfrac{8.7}{30\%}$

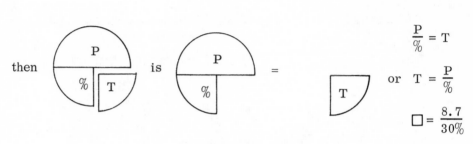

The rearranged problem is $\square = \dfrac{8.7}{30\%}$. Solve this problem.

Check your answer in **21.**

21

$\square = \dfrac{8.7}{30\%} = \dfrac{8.7}{.30}$

$\square = 29$

$$.30\overline{)8.70}$$
$$\underline{6\ 0}$$
$$2\ 70$$
$$\underline{2\ 70}$$

Guess: $\dfrac{9}{.3}$ is 30.

$$.3\overline{)9.0.0}\quad 3\ 0.$$

A reasonable guess is 30.

Check: 29 is very close to our guess.

Double-check: 30% of 29 = ?

$$.30 \times 29 = 8.7$$

𝕴𝖒𝖕𝖔𝖗𝖙𝖆𝖓𝖙 We cannot divide by 30%. We must change the percent to a decimal number before we do the division.

Here are a few practice problems to test your mental muscles.

(a) 16% of what number is equal to 5.76?

(b) 41 is 5% of what number?

(c) Find a number such that $12\frac{1}{2}\%$ of it is $26\frac{1}{4}$.

(d) 2 is 8% of a number. Find the number.

(e) 125% of what number is 35?

Check your answers against ours in **22.**

I can tell which is TOTAL and which is PART, because TOTAL is always the larger number. Right?

No! You can have a percent greater than 100%, meaning that the percentage (or PART) is more than the base (or Total). 125% of 40 is 50. 40 is the base or reference number and 50 is the percentage. **Read** the problem to tell which is which.

22

(a) $16\% \times \square = 5.76$

$$\square = \frac{5.76}{16\%}$$

$$\square = \frac{5.76}{.16}$$

$$\square = 36$$

$$\begin{array}{r} 36. \\ .16\overline{)5.76.} \\ \underline{4\ 8} \\ 96 \\ \underline{96} \end{array}$$

$T = \dfrac{P}{\%}$

Guess: $\dfrac{5.76}{.16} \cong \dfrac{500}{10}$ or about 50.

$16\% = 0.16$

Check: The guess and the answer are reasonably close.

Double-check: $16\% \times 36 = ?$

$.16 \times 36 = 5.76$

(b) $41 = 5\% \times \square$

$$\square = \frac{41}{5\%}$$

$$\square = \frac{41}{.05}$$

$$\square = 820$$

$$\begin{array}{r} 8\ 20. \\ 0.05\overline{)41.00.} \\ \underline{40} \\ 1\ 0 \\ \underline{1\ 0} \\ 0 \\ \underline{0} \end{array}$$

$T = \dfrac{P}{\%}$

Guess: $\dfrac{40}{.05} \cong \dfrac{4000}{5}$ or about 800.

Check: The guess and answer are very close.

Double-check: $5\% \times 820 = ?$

$.05 \times 820 = 41$

(c) $26\dfrac{1}{4} = 12\dfrac{1}{2}\% \times \square$

$$T = \dfrac{P}{\%}$$

or $\square = \dfrac{26\dfrac{1}{4}}{12\dfrac{1}{2}\%}$

Guess: $\dfrac{26}{.1}$ is $\dfrac{260}{1}$ or 260.

$12\dfrac{1}{2}\% = 12.5\% = 0.125$

$\square = \dfrac{26.25}{0.125}$

$$\begin{array}{r} 210. \\ 0.125\overline{)26.250.} \\ 25\ 0 \\ \hline 1\ 25 \\ 1\ 25 \\ \hline 0 \\ 0 \\ \hline \end{array}$$

$\square = 210$

Check: 210 and 260 are fairly close.

Double-check: $12\dfrac{1}{2}\% \times 210 = ?$

$.125 \times 210 = 26.25 = 26\dfrac{1}{4}$

(d) $2 = 8\% \times \square$

$$T = \dfrac{P}{\%}$$

or $\square = \dfrac{2}{8\%}$

Guess: $\dfrac{2}{.08} \cong \dfrac{2}{.1} = 20$

$\square = \dfrac{2}{.08}$

$$\begin{array}{r} 25. \\ .08\overline{)2.00.} \\ 1\ 6 \\ \hline 40 \\ 40 \\ \hline \end{array}$$

$8\% = 0.08$

$\square = 25$

Check: 20 and 25 are close enough.

Double-check: $8\% \times 25 = ?$

$.08 \times 25 = 2.00$

(e) $125\% \times \square = 35$

$$T \times \dfrac{P}{\%}$$

or $\square = \dfrac{35}{125\%}$

$125\% = 1.25$

$\square = \dfrac{35}{1.25}$

$$\begin{array}{r} 28. \\ 1.25\overline{)35.00.} \\ 25\ 0 \\ \hline 10\ 00 \\ 10\ 00 \\ \hline \end{array}$$

Guess: $\dfrac{35}{1.25}$ is less than 35. Guess about 30.

$\square = 28$

Check: The guess (30) and answer (28) agree.

Double-check: $125\% \times 28 = ?$

$1.25 \times 28 = 35$

A Review

Let's review the seven steps for solving percent problems.

- Step 1: Translate the problem sentence into a math equation.

- Step 2: Label the numbers as base or total (T), percentage or part (P), and percent (%).

- Step 3: Rearrange the math equation so that the unknown quantity is alone on the left. Use the Equation Finder.

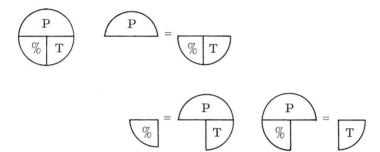

- Step 4: Guess. Get a reasonable estimate of the answer.

- Step 5: Solve the problem by doing the arithmetic. Always change percent numbers to decimal numbers first.

- Step 6: Check your answer by comparing it with the guess in step 4.

- Step 7: Double-check the answer if you can by putting it back into the original problem to see if it is correct.

Are you ready for a bit of practice on the three basic kinds of percent problems? Wind your mind and turn to **23** for a problem set.

23

Problem Set 2: Basic Percent Problems

A. 4 is _____% of 5.   15 is _____% of 75.

6% of $150 = _____   What percent of 25 is 16? _____

20% of what number is 3? _____   25% of 428 = _____

8 is what percent of 8? _____   120% of 45 is _____

35% of 60 is _____   9 is 15% of _____

8 is _____% of 64.   3% of 5,000 = _____

100% of what number is 59? _____   2.5% of what number is 2? _____

What percent of 54 is 36? _____   60 is _____% of 12.

17 is 17% of _____.   13 is what percent of 25? _____

74% of what number is 370? _____ $8\frac{1}{2}$% of $250 is _____

B. 75 is $33\frac{1}{3}$% of _____. $137\frac{1}{2}$% of 5640 is _____

What percent of 10 is 2.5? _____   21 = $116\frac{2}{3}$% of _____

6% of $3.29 is _____.   63 is _____% of 35.

12.5% of what number is 20? ___ $33\frac{1}{3}$% of $8.16 = _____

9.6 is what percent of 6.4? _____   What % of 28 is 3.5? _____

What percent of 7.5 is 2? _____ $37\frac{1}{2}$% of 12 is _____.

$.75 is _____% of $37.50.   .516 is what percent of 7.74? ___

$6\frac{1}{4}$% of 280 = _____ $2\frac{1}{4}$ is what percent of 9? _____

1.28 is _____% of .32.   42.7 is 10% of _____.

260% of 8.5 is _____.   4.75% of what number is 76? ___

C. Brain Boosters

1. If you answer 37 problems correctly on a 42-question test, what percent score do you have?

2. Fifty percent more than what number is 25% less than 32?

3. What is 40% of 90% of 140?

4. What is the difference between $\frac{1}{2}$ of 50% of 17.4 and 50% of $\frac{1}{2}$ of 17.4?

5. The population of Boomville increased from 48,200 to 63,850 in two years. What was the percent increase?

6. A marathon runner weighing a normal 146 pounds, weighed 138 after his 26-mile race. What percent of his body weight did he lose?

The answers to these problems are on page 233. When you have had the practice you need, either return to the preview on page 175 or continue in frame **24** with the study of some applications of percent to practical problems.

Applications

24

The simplest practical use of percent is in the calculation of a part (or percentage) of some total. For example, sales persons are often paid on the basis of their success at selling and receive a commission or share of the sales receipts. Commission is usually described as a percent of sales income.

Suppose your job as a door-to-door encyclopedia salesman pays 12% commission on all sales. How much do you earn from the sale of one $400 set of books?

$$12\% \text{ of } \$400 = \underline{\quad ? \quad}$$

Commission = $12\% \times \$400$

$= 0.12 \times \$400$

$= \$48$

Check: 12% is roughly $\frac{1}{10}$ and $\frac{1}{10}$ of $400 is $40.

Try this one yourself:

At the Happy Bandit Used Car Company each salesman receives a 6% commission on his sales. What would a salesman earn if he sold a 1970 Airedale for $1299.95?

Check your answer in **25.**

25

$$6\% \text{ of } \$1299.95 = \underline{\quad ? \quad}$$

Commission = $6\% \times \$1299.95$

$= 0.06 \times \$1299.95$

$= \$77.9970$ or $78.00 rounded off

Check: 10% of $1200 is $120. The answer will be about half of this or $60.

The salesman earns a commission of $78 on the sale. If this same salesman earns $255 in commissions in a given week, what was his sales total for the week?

Translate the question to a basic percent problem and solve it. Our solution is in **27.**

26

commission = percent rate × total sales cost

$$\$2000 = \square\% \times \$15,000$$

$$\square\% = \frac{\$2,000}{15,000} = \frac{2}{15}$$

$$\square = 13\frac{1}{3}\%$$

Check: 2 is more than 10% of 15. The answer is somewhere between 10 and 20%.

Ready for some practice? Try these.

(a) A real estate salesman sells a house for $34,500. His usual commission is 7%. How much does he earn on the sale?

(b) All salespeople in the Ace Junk Store receive $60 per week plus a 2% commission. If you sold $975 worth of junk in a week, what would be your income?

(c) A salesman at the Wasteland TV Company sold five color television sets last week and earned $128.70 in commissions. If his commission is 6%, what does a color TV set cost?

The correct solutions are in **28**.

27

commission = 6% × total

$255 = 6% × □   Check: 10% of what equals $250?

$$□ = \frac{\$255}{6\%} = \frac{\$255}{.06}$$

About $2500. The total
sales will be almost double
this or $4000 to $5000.

□ = $4250.00

His week's sales total was $4250.

What rate of commission would a salesman be receiving if he sold a boat for $15,000 and received a commission of $2,000? Look in **26** for the answer.

28

(a) commission = 7% × $34,500   Check: 10% of $30,000 is $3000.
He'll earn almost $3000.

= .07 × $34,500

= $2415

(b) commission = 2% × $975   Check: 2% of $100 is $2. $975
is almost 10 times this
or almost $20.

= .02 × $975

= $19.50

income = $60 + $19.50 = $79.50

(c) commission = 6% × total

$$\$128.70 = 6\% \times \square$$

$$\square = \frac{\$128.70}{6\%}$$

$$\square = \frac{\$128.70}{.06}$$

$$\square = \$2145 \text{ total sales}$$

cost per TV set = $2145 ÷ 5 or $429

Check: 6% is about $\frac{1}{16}$ and 16 × $128 is about $2000.

Turn to **29** for a look at another application of percent.

29

Another important kind of percent problem involves the idea of dis-count. In order to stimulate sales, a merchant may offer to sell some item at less than its normal price. A discount is the amount of money by which the normal price is reduced. It is a percentage or part of the normal price. Several new words are worth learning. Discount is the reduction in price. List price is the normal, regular, or original price before the discount is subtracted. Sales price is the new or discount price. The sales price is always less than the list price, of course. Discount rate is a percent number that enables you to calculate the dis-count as a part of the regular or list price.

Let's try a problem.

> The list price of a lamp is $18.50. On a special sale it is offered at 20% off. What is the sale price?

Can you solve this problem? Try, then turn to **30** for help.

30

discount = 20% of $18.50

= 0.20 × $18.50

= $3.70

Check: 20% is $\frac{1}{5}$ and $\frac{1}{5}$ of $18 is about $3. The sale price will be about $15.

sale price = list price − discount

= $18.50 − $3.70

= $14.80

Think of it this way:

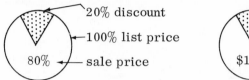

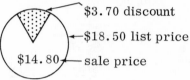

Ready for another problem?

After a 25% discount, the sale price of a camera is $144.
What was its original or list price?

Check your answer in **31.**

31

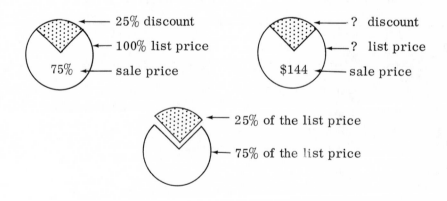

75% of list price = $144

$$75\% \times \square = \$144$$

Check: If it cost $200 list price,

$$\square = \frac{\$144}{75\%} = \frac{\$144}{.75}$$

a 25% or $\frac{1}{4}$ discount

$$\square = \$192$$

would give a sale price of $150. The list price must be about $200.

Here are a few problems to test your understanding of the idea of discount.

(a) An after-Christmas sale advertises all toys 70% off. What would be the sale price of a model space ship that cost $19.95 before the sale?

(b) A refrigerator is on sale for $376 and is advertised as "12% off regular price." What was its regular price?

(c) A set of four 740-15 automobile tires is on sale for 15% off list price. What would be the sale price if the list price is $20.80 each?

Work hard at these. Knowing how to do them may save you a lot of money one day. Check your answers in **32.**

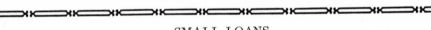

SMALL LOANS

Sooner or later everyone finds it necessary to borrow money. When you do you will want to know beforehand how it works. Suppose you borrow $200 and the loan company specifies that you repay it at $25 per month plus interest at 3% per month on the unpaid balance. What interest do you actually pay?

month 1 $200 × 0.03 = $6.00	you pay $25 + $6.00 = $31.00
month 2 $175 × 0.03 = $5.25	you pay $25 + $5.25 = $30.25
month 3 $150 × 0.03 = $4.50	you pay $25 + $4.50 = $29.50
month 4 $125 × 0.03 = $3.75	you pay $25 + $3.75 = $28.75
month 5 $100 × 0.03 = $3.00	you pay $25 + $3.00 = $28.00
month 6 $ 75 × 0.03 = $2.25	you pay $25 + $2.25 = $27.25
month 7 $ 50 × 0.03 = $1.50	you pay $25 + $1.50 = $26.50
month 8 $ 25 × 0.03 = $ .75	you pay $25 + $0.75 = $25.75
$27.00	$227.00
total interest	total of 8 loan
is $27.00	payments

The loan company might also set the loan up as 8 equal payments of $227.00 ÷ 8 = $28.375 or $28.38 per month.

The 3% per month interest rate seems small but it amounts to about 20% per year.

A bank loan for $200 at 7% for 8 months would cost you:

$$\$200 \times 7\% \times \frac{8}{12}$$

$$\text{or}\quad \$200 \times .07 \times \frac{8}{12}$$

or $9.34 (Quite a difference compared to $27.)

The loan company demands that you pay more and in return they are less worried about your ability to meet the payments. For a bigger risk, they want a higher interest.

32

(a) discount = 70% of $19.95

$$= .70 \times \$19.95$$

$$= \$13.985 \text{ or } \$13.99 \text{ (rounded)}$$

sale price = list price − discount

$$= \$19.95 - \$13.99$$

$$= \$5.96$$

(b) discount = 12% of list price
sale price = 88% of list price

$$\$376 = 88\% \times \square$$

$$\square = \frac{\$376}{88\%} = \frac{\$376}{.88}$$

$$\square = \$427.28 \text{ (rounded)}$$

(c) discount = 15% of list price

$$= 15\% \times (4 \times \$20.80)$$

$$= .15 \times \$83.20$$

$$= \$12.48$$

sale price = list price − discount

$$= \$83.20 - \$12.48$$

$$= \$70.72$$

Taxes are almost always calculated as a percent of some total amount. Property taxes are written as some fraction of the value of the property involved. Income taxes are most often calculated from complex formulas that depend on many factors. We cannot consider either income or property taxes here.

A sales tax is an amount calculated from the actual price of a purchase and added to the buyer's cost. Retail sales tax rates are set by the individual states in the United States and vary from 0 to 7% of the sales price. A sales tax of 6% is often stated as "6¢ on the dollar" since 6% of $1.00 equals 6¢.

Here is an example of a typical tax problem:

> If the retail sales tax rate is 6% in California, how much would you pay in Los Angeles for a pair of shoes costing $16.50?

Try it, then check your work in **33**.

33

sales tax = 6% of $16.50

= 0.06 × $16.50

= $0.99

Check: 6¢ on each dollar; $16; then the sales tax should be about 6 × 16 or 96¢.

actual cost = list price + sales tax

= $16.50 + $0.99

= $17.49

Most stores and sales clerks use tables to look up the sales tax and therefore do not need to do the arithmetic shown above except on large purchases beyond the range of the tables. However, it is your best interest to be able to check their work.

Here are a few problems to test your ability to calculate sales tax. If sales tax is 5% find the sales tax on each of the following:

(a) a pen priced at 39¢ _____

(b) a chair priced at $27.50 _____

(c) a toy priced at $2.95 _____

(d) a new car priced at $2785 _____

(e) a bicycle priced at $126.50 _____

(f) a tube of toothpaste priced at 79¢ _____

Check your answers in **34**.

LIVING AND DYING ON THE INSTALLMENT PLAN

The biggest financial deal most people ever undertake is the purchase of a car. Suppose you buy a new car for \$3600, pay one-sixth down, and obtain a loan from the finance company for the remainder at $7\frac{1}{2}\%$ to be paid in 24 monthly payments.

down payment = \$3600 ÷ 6 = \$600

loan = \$3600 − \$600 = \$3000

interest = \$3000 × $7\frac{1}{2}\%$ × 2

          amount of   interest   time in years
          the loan   rate per year

interest = \$3000 × 0.075 × 2

       = \$450

total to be paid = \$3000 + \$450 = \$3450

payments = \$3450 ÷ 24 = \$143.75

You pay \$18.75 (\$450 ÷ 24 = \$18.25) each month for 24 months to use the "easy payment plan" and it may be worth it in order to have the car to use now rather than wait until you can save up the money to buy it with cash.

Repeat the calculation above with a down payment of only \$100 and see what a large difference in total cost that makes.

34

	tax	total cost	calculation
(a)	2¢	41¢	.05 × 39¢ = \$1.95 ≅ 2¢
(b)	\$1.38	\$28.88	.05 × \$27.50 = \$1.375 ≅ \$1.38
(c)	\$0.15	\$3.10	.05 × \$2.95 = \$0.1475 ≅ \$0.15
(d)	\$139.25	\$2924.25	.05 × \$2785 = \$139.25
(e)	\$6.33	\$132.83	.05 × \$126.50 = \$6.325 ≅ \$6.33
(f)	4¢	83¢	.05 × 79¢ = \$3.95 ≅ 4¢

In our modern society we have set up complex procedures that allow you to use someone else's money. A <u>lender</u> with money beyond his needs supplies cash to a <u>borrower</u> whose needs exceed his money. The money is called a loan. <u>Interest</u> is the amount the lender is paid for the use of his money. Interest is the money you pay to use someone else's money. The more money you borrow and the longer you keep it, the more interest you must pay. When you purchase a house with a bank loan, a car or a refrigerator on an installment loan, or gasoline on a credit card, you are using someone else's money and you pay interest for that use. If you are on the other end of the money game, you may earn interest for money you invest in a savings account or in shares of a business.

Suppose you have $500 in savings earning $5\frac{1}{4}\%$ annually in your local bank. How much interest do you receive in a year? Compute the interest this way:

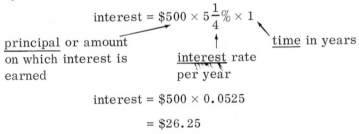

$$\text{interest} = \$500 \times 5\frac{1}{4}\% \times 1$$

principal or amount on which interest is earned   interest rate per year   <u>time</u> in years

$$\text{interest} = \$500 \times 0.0525$$

$$= \$26.25$$

By placing your $500 in a bank savings account you allow the bank to use it in various profitable ways and you are paid $26.25 per year for its use.

Most of us play the money game from the other side of the counter. Suppose you find yourself in need of cash and arrange to obtain a loan from a bank. You borrow $600 at 8% per year for 3 months. How much interest must you pay?

Try to set up and solve this problem exactly as we did in the problem above. Our worked solution is in **35**.

HOW DOES A CREDIT CARD WORK?

Many forms of small loans, such as credit card loans, charge interest by the month. These are known as "revolving credit" plans. (If you buy very much money this way, you do the revolving and may run in circles for years trying to pay it back!) Essentially, you buy now and pay later. If you repay the full amount borrowed within 25 or 30 days there is no charge for the loan.

After the first pay period of 25 or 30 days, you pay a percent of the unpaid balance each month, usually between one-half and 2%. In addition you must pay some minimum amount each month, usually $10 or 10% of the unpaid balance, whichever is larger. Generally you are also charged a small monthly amount for insurance premiums. (The credit card company insures themselves against your defaulting on the loan or disappearing, and you pay for their insurance.)

Let's see how it works. Suppose you go on a short vacation and pay for gasoline, lodging, and meals with your Handy Dandy credit card. A few weeks later you receive a bill for $100. You can pay it within 30 days and owe no interest or you can pay over several months as follows:

month 1 $100 \times 1\frac{1}{2}\% = \$100 \times 0.015 = \$1.50$ owe \$101.50

           pay \$10 and carry \$91.50 over to next month

month 2 $\$91.50 \times 0.015 = \$1.38$ owe \$92.88

           pay \$10 and carry \$82.88 over to next month

. . . and so on.

A year later you will have repaid the $100 loan and all interest.

The $1\frac{1}{2}\%$ per month interest rate seems small, but it is equivalent to between 15% and 18%. You pay at a high rate for the convenience of using the credit card and the no-questions-asked ease of getting the loan.

35

$$\text{interest} = \$600 \times 8\% \times \frac{3}{12} \leftarrow \text{time in years} \quad \frac{3 \text{ months}}{12 \text{ months}} = \frac{3}{12} \text{ year}$$

principal or   interest rate
amount loaned   per year

$$\text{interest} = \$600 \times 0.08 \times \frac{3}{12}$$

$$= \$12$$

Depending on how you and the bank decide to arrange it, you may be required to pay the total principal ($600) plus interest ($12) all at once at the end of three months, or you may use some sort of regular payment plan (for example, pay $204 each month for three months).

What interest do you pay if you borrow $1200 for two years at $8\frac{1}{2}\%$ interest per year? The answer is in **36**.

36

$$\text{interest} = \$1200 \times 8\frac{1}{2}\% \times 2 \text{ years}$$

$$= \$1200 \times 0.085 \times 2$$

$$= \$204$$

You repay $1200 + $204 = $1404 over two years, perhaps in 24 monthly payments of $1404 ÷ 24 or $58.50 each.

Now that we have finished our very brief excursion into the mysteries of high finance, turn to **37** for a set of practice problems on these important concepts.

37

Problem Set 3: Practical Applications of Percent

1. A lawyer collected $8500 for a client on a damage suit and he charged $2480 for his services. What was his rate of commission?

2. How much money must you invest at 6% to earn $3000 in a year?

3. A local sporting goods store offers coaches a 20% discount on all merchandise. The cricket coach at Madam MacAdam's Academy bought a new wicket for $26.96. What was the discount price he paid?

4. A typewriter sells for $176 after a 12% discount. What was its original or list price?

5. If the retail sales tax in your state is 4%, what would be the total cost of each of the following:

   (a) a $4500 sports car
   (b) a $1.98 toy
   (c) 69¢ worth of nails
   (d) a $10.60 textbook
   (e) a $3.10 picture frame

6. Students selling tickets to the town carnival received 40¢ for every $2.50 ticket sold. What percent commission is this?

7. A refrigerator has a list price of $329. You buy it while it is on sale at 15% discount and you agree to pay $40 down and the remainder in 12 equal monthly payments at 12% interest per year.

   (a) What is the sale price of the refrigerator?
   (b) What total interest will you pay?
   (c) How big will your payments be?

8. What is the selling price of a set of golf clubs with a list price of $165 if they are on sale at a 35% discount?

9. Sam wants to buy a new car costing $4000. Being very experienced in money matters he visits several banks, shopping for the best loan. At the First National Bank they offer to finance $3900 at only 7% interest over 3 years. At the Last National Bank they offer to finance only 80% of his car and they want 8% interest over 3 years. Which is the better loan? Calculate the interest for each.

10. In a newspaper advertisement a bicycle is offered for sale: "Save $14.70, buy now at 12% off the regular price." What was the regular price?

The answers to these problems are on page 233. When you have had the practice you need turn to **38** for a self-test on percent problems.

38

Chapter 4 Self-Test

1. Write $3\frac{1}{6}$ as a percent. _____

2. Write $\frac{5}{12}$ as a percent. _____

3. Change 0.08 to a percent. _____

4. Write 6.43 as a percent. _____

5. Write 2% as a decimal. _____

6. Write $112\frac{1}{2}$% as a decimal. _____

7. Write 68% as a fraction. _____

8. Find 48% of 250. _____

9. Find 63% of 12. _____

10. Find 165% of 70. _____

11. Find $6\frac{1}{2}$% of 134. _____

12. Find 46.5% of 13.4. _____

13. What percent of 26 is 9.1? _____

14. What percent of 75 is 112.5? _____

15. What percent of 0.72 is 1.62? _____

16. What percent of 1.45 is 0.609? _____

17. What percent of $\frac{3}{8}$ is $\frac{1}{9}$? _____

18. 15% of what number is 12? _____

19. If 120% of a number is 0.85, find the number. (Round to two decimal digits.) _____

20. A gallon of gas costs 62.9¢. If the price is increased 20% find the new cost. _____

21. After a 30% discount an article costs $43.40. Find the original price. _____

22. What commission does a salesman earn on a $370 sale if his commission rate is 16%? _____

23. What is the sale price of a book marked 40% off if its list price is $8.95? _____

24. What is the interest on a $400 loan at $7\frac{1}{4}\%$ over 30 months?

————————

25. If the retail sales tax is 6%, what would be the total cost of a $1.85 toy? ——————

The answers to these problems are on page 235.

Final Exam

1. $65 + 284 =$ _____

2. $7852 + 519 + 604 =$ _____

3. $403 - 186 =$ _____

4. $7201 - 5325 =$ _____

5. $58 \times 27 =$ _____

6. $354 \times 806 =$ _____

7. $1128 \div 24 =$ _____

8. $38\overline{)17366} =$ _____

9. Factor: $378 =$ _____

10. Factor: $1540 =$ _____

11. $\sqrt{256} =$ _____

12. Write $8\frac{2}{9}$ as an improper fraction: _____

13. Write $\frac{47}{6}$ as a mixed number: _____

14. Reduce to lowest terms: $\frac{660}{924} =$ _____

15. $\frac{3}{4} + \frac{2}{5} =$ _____

16. $1\frac{7}{8} + 2\frac{3}{5} =$ _____

17. $2\frac{1}{4} + 4\frac{5}{6} + 2\frac{2}{3} =$ _____

18. $\frac{3}{5} - \frac{1}{3} =$ _____

19. $4\frac{1}{2} - 2\frac{3}{7} =$ _____

20. $\frac{5}{6} \times \frac{3}{4} =$ _____

21. $2\frac{4}{7} \times \frac{5}{6} =$ _____

22. $\dfrac{3}{8} \div \dfrac{9}{10} =$ _____

23. $3\dfrac{1}{5} \div 2\dfrac{4}{7} =$ _____

24. $(4\dfrac{2}{3})^2 =$ _____

25. What fraction of $5\dfrac{2}{3}$ is $1\dfrac{1}{4}$? _____

26. $11.036 + 7.8 =$ _____

27. $129.4 + 6.77 + 28.025 =$ _____

28. $32 - 6.43 =$ _____

29. $8 - 0.27 =$ _____

30. $14.21 - 1.962 =$ _____

31. $9.4 \times 4.3 =$ _____

32. $.705 \times 48.4 =$ _____

33. $6 \div 9.4$ (round to two decimal digits) $=$ _____

34. $0.0564 \div .03 =$ _____

35. Write 4.276 as a fraction in lowest terms: _____

36. Write $\dfrac{5}{6}$ as a decimal: _____

37. What part of 1.7 is 1.36? _____

38. Write $2\dfrac{3}{8}$ as a percent: _____

39. Write 0.01 as a percent: _____

40. Write 128% as a decimal: _____

41. Find 6% of 450. _____

42. Find 125% of 5.6. _____

43. What percent of 56 is 7? _____

44. 32% of what number is 1.6? _____

45. What percent of $\dfrac{2}{3}$ is $\dfrac{1}{5}$? _____

46. At the start of a vacation trip the car odometer reads 23,463 miles. Two weeks later at the end of the trip it reads 27,205 miles. How many miles were traveled on the trip?

47. If sound travels at the rate of 1088 feet per second, how far does it travel in one-half of an hour?

48. If $851 were divided among 23 people, how much money would each person receive?

49. A $3\frac{3}{4}$ pound package of soy beans costs 75 cents. What should 2 pounds cost?

50. Sue carefully counted out the coins in her purse. She had 7 quarters, 12 dimes, 15 nickels, and 29 pennies. How much money did she have in coins?

51. A camera sells for $225 after a 25% discount. What was its original price?

52. A book normally sells for $7.95 and is on sale at a 20% discount. What will it cost, including 6% sales tax?

53. What is the cost of 9.7 gallons of gasoline at 57.9 cents per gallon?

54. What fraction of 60 is $8\frac{1}{2}$?

55. How much money must you have deposited in savings at $5\frac{1}{2}$% in order to earn $137.50 in one year?

Appendix

Answers

Chapter 1   The Arithmetic of Whole Numbers

Problem Set 1, pages 10-11

A. 8, 16, 6, 13, 11, 11, 13, 9, 13, 12
    14, 17, 6, 12, 15, 9, 11, 12, 16, 9
    15, 10, 10, 11, 14, 7, 8, 14, 18, 15
    14, 14, 12, 12, 16, 12, 13, 13, 13, 15
    11, 17, 11, 12, 11, 15, 13, 12, 18, 12

B. 10, 11, 11, 12, 16, 9, 9, 12, 12, 13
    10, 13, 15, 8, 14, 9, 13, 18, 15, 11
    11, 16, 14, 10, 17, 9, 14, 12, 7, 10
    13, 13, 12, 16, 9, 11, 10, 10, 14, 17
    11, 14, 11, 10, 11, 14, 16, 12, 13, 15

C. 11, 12, 14, 17, 18, 21, 11, 20, 18, 15
    15, 14, 15, 18, 22, 17, 14, 14, 18, 16
    12, 19, 8, 10, 22, 12, 19, 15, 14, 16

Box, page 16

91, 81, 46, 92, 74, 140

Problem Set 2, pages 17-20

A. 70, 104, 65, 126, 80, 106
    112, 72, 131, 103, 105, 123
    103, 124, 100, 132, 52, 136

B. 415, 393, 1113, 1003, 1530
    1390, 1016, 831, 1262, 1009
    824, 806, 1241, 861, 1001

Problem Set 2, pages 17-20 (continued)

C. 5525, 9563, 9461, 2611   D. 25717, 11071, 70251, 21642
    9302, 3513, 3702, 12599   14711, 89211, 47111
    7365, 10122, 6505, 11428   175728, 101011, 180197
    5781, 15715, 9403, 11850

E. 1042, 5211, 2442, 6441, 7083, 16275, 6352, 7655, 6514, 9851

F. 1. 882 lb   2. 371 calories   3. 662,289 miles
    4. $10,535   5. They are the same: 1,083,676,269
    6. 8454 points

Problem Set 3, pages 27-28

A. 6, 5, 7, 6, 2, 5, 8, 0, 4, 1
    9, 8, 3, 7, 0, 9, 3, 7, 3, 4
    8, 1, 8, 4, 9, 7, 9, 7, 9, 4
    9, 6, 3, 1, 6, 6, 8, 5, 5, 8
    7, 7, 18, 7, 7, 3, 8, 9, 0, 2

B. 13, 44, 29, 16, 12, 57, 19, 17, 15
    28, 36, 18, 22, 37, 25, 26, 38, 85

C. 189, 458, 85, 877, 281, 176, 154, 266
    273, 198, 715, 51, 574, 45, 29, 145

D. 2809, 7781, 5698, 28842, 12518, 7679
    56042, 37328, 4741, 9897, 9614, 26807
    47593, 316640, 22422, 55459, 24939

E. 1. 1819   2. 284   3. 13,819
    4. $7289   5. $155   6. 3,303
    7. 3   8. $595
    9. 98 - 76 + 54 + 3 + 21   10. Sure, 1,963 pennies are
        123 + 45 - 67 + 8 - 9   worth $19.63, in fact.
        12 + 3 + 4 + 5 - 6 - 7 + 89
        123 + 4 - 5 + 67 - 89

Problem Set 4, pages 31-32

A. 12, 32, 63, 36, 12, 18, 0, 24, 14, 8
    48, 16, 45, 30, 10, 9, 72, 35, 18, 0
    28, 15, 36, 49, 8, 40, 42, 54, 64, 24
    20, 0, 25, 27, 81, 6, 1, 48, 16, 63

B. 16, 30, 9, 35, 18, 20, 28, 48, 12, 63
    32, 0, 18, 24, 9, 25, 24, 45, 10, 72
    15, 49, 40, 54, 36, 8, 42, 64, 0, 4
    25, 27, 7, 56, 36, 12, 81, 0, 2, 56

Problem Set 5, pages 40–41

A. 42, 72, 56, 63, 48, 54
    72, 42, 54, 63, 56, 48

B. 87, 402, 576, 243, 282, 792, 320, 259, 156, 294
    290, 261, 564, 392, 153, 161, 282, 424, 308, 324
    720, 1505, 1728, 2736, 5040, 2952, 7138, 1170, 1938, 2548
    1650, 1349, 4425, 1458, 928, 6232, 3822, 2030, 8930, 2752

C. 37515, 74820, 375750, 97643, 297591, 384030, 38023
    108486, 378012, 1279840, 41064, 4947973, 30780, 225852
    1368810, 31152, 397584, 43381, 60241, 5098335

D. 1. $338   2. 8760   3. $1196
    4. $205   5. 378   6. $130640
    7. 987654321
    8. 111   1111   11111   111111   1111111   11111111
    9. 123458769   123547689   967854321

Box, page 43

1. 3 hours   2. 84.5 points   3. $663.45

Problem Set 6, pages 49–51

A. 9, 12, 11 R4   B. 35, 41, 23 R6
    7 R2, not defined, 13   42, 57, 46 R2
    7, 5, 10 R1   51 R4, 45, 112 R1
    7, 1, not defined   44, 21, 27
    8, 6, 4   52, 37, 88
    7, 6, 9   125, 50 R1, 67

C. 23, 21, 20 R2   D. 95 R6, 104, 96
    31 R4, 39, 50 R2   208, 142 R6, 107
    25, 19 R17, 9 R1   222 R2, 171, 32
    41, 53, 43   1000, 305 R5, 311 R8
    22, 11 R34, 12   84 R41, 100 R5, 119
    34, 9 R6, 71 R5   61, 102 R98, 81

E. 1. 7 mph   2. $507
    3. 207 pounds   4. 38   5. $51
    6. 25 minutes   7. 247 years, 309 days
    8. 344 meters per second

Problem Set 7, pages 63–65

A. $2 \times 2 \times 3$ $2 \times 2 \times 2 \times 2$ 2×7

$2 \times 3 \times 3$ $2 \times 2 \times 2 \times 3$ $2 \times 2 \times 5$

2×13 prime $2 \times 2 \times 2 \times 2 \times 2$

$2 \times 2 \times 3 \times 3$ 3×13 $2 \times 3 \times 7$

$2 \times 2 \times 2 \times 7$ $3 \times 3 \times 3 \times 3$ 11×11

B. $2 \times 2 \times 2 \times 2 \times 2 \times 3$ $2 \times 2 \times 3 \times 7$ $2 \times 2 \times 2 \times 17$

$2 \times 5 \times 17$ $2 \times 2 \times 3 \times 3 \times 7$ $2 \times 2 \times 2 \times 2 \times 2 \times$
                                                         $2 \times 2 \times 2$

$2 \times 2 \times 2 \times 2 \times 2$ $2 \times 3 \times 5 \times 13$ $2 \times 2 \times 3 \times 3 \times 13$
           3×3

$2 \times 3 \times 7 \times 13$ $2 \times 2 \times 5 \times 7 \times 7$ 37×37

29×47 $2 \times 2 \times 3 \times 3 \times 43$ 47×67

C. primes: 2, 5, 3, 31, 23, 37, 53, 19, 67, 61, 89, 17

D. divisible by 2: 12, 4, 144, 1044, 1390, 72, 102, 2808, 2088, 8280, 8802

divisible by 3: 9, 12, 231, 45, 144, 261, 1044, 72, 81, 102, 2808, 2088, 8280, 8802, 111

divisible by 5: 45, 1390, 8280

E. 1. $28 = 1 + 2 + 4 + 7 + 14$

$496 = 1 + 2 + 4 + 8 + 16 + 31 + 62 + 124 + 248$

$8128 = 1 + 2 + 4 + 8 + 16 + 32 + 64 + 127 + 254 + 508 + 1016 +$
                                       $2032 + 4064$

2. a. The divisors of 220 = 284 and the divisors of 284 = 220.

     $1 + 2 + 4 + 5 + 10 + 11 + 20 + 22 + 44 + 55 + 11 = 284$

b. $2924 = 1 + 2 + 4 + 5 + 10 + 20 + 131 + 262 + 524 + 655 + 1310$

     $2620 = 1 + 2 + 4 + 17 + 34 + 43 + 68 + 86 + 172 + 731 + 1462$

3. 37; all numbers are primes

4. 2, 3, 6

5. 1, not prime; 11, prime; $111 = 3 \times 37$; $1111 = 11 \times 101$;

$11111 = 41 \times 271$; $111111 = 3 \times 37 \times 1001$

6.   638   4752   314

        475   3658   926

    + 253   4975   + 705

     1366   + 2403   1945

                    15788

Problem Set 8, pages 72–74

A. 16, 9, 64, 125

1000, 49, 256, 36

512, 81, 625, 100,000

8, 243, 729, 1

6, 1, 256, 32

Problem Set 8, pages 72–74   (continued)

A.   1,000,000, 343, 64, 1296
     1024, 6561, 216, 25
     27, 2401, 1, 10,000
     16, 5, 4096, 1
     81, 1024, 100, 7776

B.   196, 441, 3375
     15,625, 256, 3025
     3721, 64,000, 1000000
     108, 576, 1125
     7938, 4851, 2025
     2744, 1296, 24300
     2000, 90,000, 9216

C.   9, 12, 4, 5
     6, 10, 7, 18
     1, 11, 8, 3
     15, 2, 20, 16

D.   3.   (a) $5^1 = 5$ $5^2 = 25$ $5^3 = 125$ $5^4 = 625$
          (b) $25^1 = 25$ $25^2 = 625$ $25^3 = 15625$ $25^4 = 390625$
          (c) $625^1 = 625$ $625^2 = 390625$ $625^3 = 244140625$
               $625^4 = 152,587,890,625$
          (d) $376^1 = 376$ $376^2 = 141376$ $376^3 = 53157376$
               $376^4 = 19987173376$

Chapter 2   Fractions

Problem Set 1, pages 90–92

A. $\dfrac{7}{3}$, $\dfrac{22}{5}$, $\dfrac{15}{2}$, $\dfrac{94}{7}$, $\dfrac{35}{4}$

     $\dfrac{4}{1}$, $\dfrac{5}{3}$, $\dfrac{35}{6}$, $\dfrac{31}{8}$, $\dfrac{13}{5}$

     $\dfrac{161}{10}$, $\dfrac{635}{9}$, $\dfrac{481}{40}$, $\dfrac{170}{11}$, $\dfrac{113}{3}$

B. $8\dfrac{1}{2}$, $7\dfrac{2}{3}$, $1\dfrac{3}{5}$, $4\dfrac{3}{4}$, $6\dfrac{1}{6}$, $9\dfrac{1}{3}$, $4\dfrac{5}{8}$, $4\dfrac{1}{7}$

     $1\dfrac{9}{25}$, $5\dfrac{2}{9}$, $52\dfrac{3}{4}$, $7\dfrac{9}{23}$, $4\dfrac{3}{10}$, $20\dfrac{5}{6}$, $9\dfrac{4}{15}$

C. $\dfrac{13}{15}$, $\dfrac{4}{5}$, $\dfrac{4}{5}$, $\dfrac{1}{2}$, $\dfrac{1}{8}$

     $\dfrac{2}{5}$, $\dfrac{1}{6}$, $\dfrac{8}{9}$, $\dfrac{1}{3}$, $\dfrac{3}{8}$

     $\dfrac{7}{20}$, $\dfrac{3}{8}$, $\dfrac{1}{6}$, $\dfrac{4}{7}$, $\dfrac{5}{9}$

D. $\dfrac{14}{16}$, $\dfrac{27}{45}$, $\dfrac{9}{12}$, $\dfrac{29}{60}$, $\dfrac{7}{63}$

     $\dfrac{45}{35}$, $\dfrac{20}{32}$, $\dfrac{140}{25}$, $\dfrac{39}{78}$, $\dfrac{34}{51}$

     $\dfrac{363}{44}$, $\dfrac{82}{14}$, $\dfrac{66}{72}$, $\dfrac{185}{50}$, $\dfrac{516}{54}$

E.   1.   5 laps where each lap is $\dfrac{1}{8}$ of a mile

Problem Set 1, pages 90-92 (continued)

E. 3. Sugar Glops   4. too little
    5. 100

Problem Set 2, pages 96-98

A. $\dfrac{1}{8}, \dfrac{1}{9}, \dfrac{4}{15}$

$\dfrac{1}{8}, \dfrac{2}{15}, \dfrac{5}{6}$

$3, \dfrac{1}{2}, 2\dfrac{2}{3}$

$\dfrac{5}{6}, \dfrac{11}{45}, \dfrac{9}{56}$

$1\dfrac{1}{9}, 10\dfrac{1}{2}, \dfrac{13}{16}$

$\dfrac{4}{5}, 2\dfrac{1}{2}, 6$

$14, 1\dfrac{3}{7}, 1\dfrac{1}{3}$

C. $\dfrac{4}{9}, \dfrac{1}{16}, \dfrac{27}{125}$

$10\dfrac{6}{25}, 91\dfrac{1}{8}, \dfrac{3}{4}$

$\dfrac{2}{7}, \dfrac{4}{5}, \dfrac{5}{8}$

$\dfrac{9}{11}, \dfrac{1}{15}$

$\dfrac{1}{6}, 1\dfrac{1}{3}$

$20, 31\dfrac{1}{2}$

B. 3, 4

$8, \dfrac{21}{16}$

$3\dfrac{1}{4}, 62$

$1\dfrac{1}{21}, \dfrac{1}{3}$

$69, 6$

$35\dfrac{3}{4}, 1\dfrac{3}{11}$

$74, 7\dfrac{9}{10}$

$9\dfrac{7}{8}, 46\dfrac{2}{3}$

$10\dfrac{3}{8}, 13\dfrac{13}{30}$

$21\dfrac{1}{3}, 6$

D. 1. 1530 miles   2. Bert ate $\dfrac{1}{2}$ pie
    3. 110 km   4. 54¢

    6. $\dfrac{7}{8}$ square miles   7. $16\dfrac{1}{2}$ mgm

Box, page 104

1. 22   2. 462   3. 198

Problem Set 3, pages 105–107

A. $1\frac{2}{3}$, $1\frac{3}{4}$,   9

    $\frac{1}{12}$, $\frac{5}{16}$, $\frac{4}{9}$

    24, $\frac{7}{16}$, $\frac{8}{13}$

    $7\frac{1}{2}$,   1, $1\frac{1}{2}$

    $\frac{5}{28}$, $1\frac{4}{5}$, $2\frac{2}{5}$

B.   9, $7\frac{1}{3}$,   4

    $\frac{3}{4}$, $1\frac{1}{3}$,   6

    $1\frac{3}{4}$, $2\frac{4}{7}$, $1\frac{1}{5}$

    $8\frac{1}{3}$, $1\frac{1}{4}$, $1\frac{38}{57}$

    $\frac{6}{7}$, $5\frac{1}{4}$, $\frac{4}{5}$

C.   16, $\frac{3}{8}$, $\frac{1}{9}$

    $3\frac{1}{9}$,   18,   25

    $\frac{6}{7}$,   6

    $1\frac{1}{4}$, $\frac{6}{29}$

    $17\frac{1}{2}$,   17

D.   1. $10\frac{8}{13}$

    2. $4\frac{1}{2}$

    3.   34 mph

    4. $1\frac{3}{8}$

Problem Set 4, pages 118–120

A. $1\frac{2}{5}$, $1\frac{1}{3}$,   1

    $\frac{2}{3}$, $\frac{1}{4}$, $\frac{1}{2}$

    $\frac{5}{12}$, $\frac{3}{8}$, $1\frac{1}{8}$

    $1\frac{1}{8}$, $\frac{3}{4}$, $\frac{1}{4}$

    $1\frac{3}{8}$, $1\frac{2}{3}$

    $1\frac{4}{7}$, $\frac{5}{8}$

    $1\frac{1}{4}$, $\frac{9}{20}$

B. $1\frac{5}{8}$, $\frac{1}{8}$, $1\frac{11}{36}$

    $\frac{19}{36}$, $\frac{29}{48}$, $\frac{29}{35}$

    $\frac{53}{192}$, $\frac{215}{216}$, $\frac{13}{48}$

    $\frac{4}{35}$, $\frac{5}{96}$, $\frac{83}{216}$

    $1\frac{5}{8}$, $4\frac{1}{36}$, $5\frac{17}{48}$

    $4\frac{51}{56}$, $\frac{3}{4}$, $1\frac{23}{48}$

    $1\frac{23}{48}$

Problem Set 4, pages 118–120 (continued)

C. $1\frac{2}{3}$, $1\frac{13}{16}$

   $5\frac{1}{4}$, $1\frac{4}{5}$

   $20\frac{3}{4}$, $1\frac{89}{120}$

   $\frac{33}{40}$, $1\frac{17}{60}$

   $3\frac{1}{12}$

   $2\frac{13}{40}$

   $5\frac{11}{48}$

   $1\frac{9}{10}$

D. 1. $\frac{15}{56}$ 2. $36\frac{2}{3}$ hours

   4. $\frac{7}{8} = \frac{1}{2} + \frac{1}{4} + \frac{1}{8}$

      $\frac{5}{9} = \frac{1}{3} + \frac{1}{6} + \frac{1}{18}$

      $\frac{5}{12} = \frac{1}{3} + \frac{1}{12}$

   5. $414\frac{7}{8}$ ft

   6. $28\frac{2}{5}$; 65 miles per gallon

   7. $2\frac{3}{16}$

   8. $\frac{1}{4}$ minus $\frac{1}{4}$ of $\frac{1}{4}$ is larger

Problem Set 5, page 131

A. (a) =

   (b) ×

   (c) +

   (d) =

   (e) □ – 6

   (f) $\frac{1}{2} \times$ □

   (g) 2 × □

   (h) □ ÷ $2\frac{1}{2}$

   (i) □ + $\frac{2}{3}$

   (j) □ + $\frac{2}{5}$

   (k) □ ÷ $\frac{3}{7}$

   (l) $\frac{7}{8} \times$ □ $= 1\frac{1}{2}$

   (m) □ – $1\frac{3}{4}$

   (n) □ $\times 3\frac{1}{4} = 11\frac{1}{2}$

   (o) $\frac{7}{16} \times 3\frac{5}{8}$

B. (a) $1\frac{1}{2}$

   (b) $\frac{3}{10}$

   (c) $1\frac{3}{5}$

   (d) $\frac{16}{21}$

   (e) $1\frac{3}{8}$

   (f) $1\frac{1}{4}$

   (g) $14\frac{2}{5}$

C. (a) 52¢

   (b) $8\frac{1}{8}$ miles

   (c) $5\frac{1}{3}$ gallons

   (d) $4\frac{1}{8}$

   (e) $186

Chapter 3   Decimals

Problem Set 1, pages 145-147

A. 0.8, 1.6
    1.5, 0.6
    0.9, 1.2
    1.6, 0.7
    1.7, 0.1
    0.7, 3.3
    0.5, 2.3
    1.8, 1.4
    3.3, 5.2
    1.9, 9.9
    4.8, 1.4
    9.0, 18.1
    5.5, 1.4
    17.5, 1.6
    6.7, 0.6
    3.6, 1.3
    1.8, 2.6
    2.6, 0.4

B. 21.01, $15.02
    1.617, 27.19
    $30.60, 6.486
    78.17, $151.11
    5.916, 828.60
    16.2019, 1031.28
    63.7313, 238.24
    128.3685
    45.195
    $27.59
    70.871
    108.37
    19.37, $15.36
    51.34, 1.04
    3.86, 42.33
    6.63, 6.52
    6.42, $36.18
    22.016, 2.897
    $17.65, 6.96
    0.3759

C. 151.461, 602.654
    95.888, 91.15
    316.765, 14.67755
    16.0425, 4035.4933
    19.011, 3.34974

D. 1. $1.84
    2. 4687.8
    3. $13212.00
    4. 2.556 yards
    5. 2.267 inches
    6. .163826 seconds
    7. $415.35
    8. 968.749 meters

Problem Set 2, pages 160-163

A. 0.00001, 0.1
    21.5, 0.06
    0.008, 0.014
    0.09, 0.72
    0.84, 0.009
    0.00006, 4.20
    1.4743, 2.18225
    0.5022, 0.03
    .024, 2.16
    0.01476, 3.6225
    1.44, 80.35
    0.00117, 287.5
    .03265, 0.0009255
    1223.6

B. 1300, 12.6
    0.045, 13
    126, 450
    60, 2000
    10000, .037
    11.200909, 6.6
    1900, 3256.25
    0.11

Problem Set 2, pages 160-163 (continued)

C. (a) 0.33, 1.43
     0.83, 0.09
     10.53, 0.67
     37.50, 0.89
     0.12, 73.40
     2.62, 0.12
     33.86, 474.44
     5.0

   (b) 33.3, 1.1
     0.2, 6.7
     0.2, 0.1
     11.1, 285.7
     0.15, 16.0

   (c) 14.286, 0.023
     0.225, 65.00
     3.462, 13.640
     3.815, 2.999
     571.428, 1109.000

D. .09, .0009, .000009
    .027, 1.44, 1.728
    .01, .0001, .000027

$$\frac{1}{81} = .012 \text{ (rounded)}$$

$$\frac{1}{7} = .143 \text{ (rounded)}$$

    0.22, .02
    0.21, 27777.78

E. 1. $84.15
   2. $7.86
   3. $126.00
   4. $26.35
   5. 2872.148
   6. 1st: (b) $12 \div .03 = 400$
       2nd: (e) $0.2 \times 0.2 = 0.04$
       3rd: The third error is that
            there are only 2 errors.
   7. December 31
   8. Yes. It is also true for any
      other digit.

Problem Set 3, pages 172-173

A. .50, .33, .67
   .25, .50, .75
   .20, .40, .6o
   .80, .17, .83
   .14, .28, .43
   .13, .38, .63
   .88, .10, .20
   .30, .08, .17
   .25, .4, .58
   .92, .06, .19
   .31, .44, .56
   .69, .81, .94

C. 1. 1.2
   2. 408
   3. 4.5
   4. 17
   5. .5

D. 1. 4.385, 1.77
     1.475, 0.7681
     7.875, 2.296
     1.434, 0.3125

B. $\dfrac{3}{10}, \dfrac{3}{4}, \dfrac{11}{25}$

   $\dfrac{4}{5}, \dfrac{3}{5}, \dfrac{1}{40}$

   $\dfrac{2}{5}, 1\dfrac{3}{10}, 2\dfrac{1}{4}$

   $2\dfrac{1}{20}, 3\dfrac{4}{25}, 1\dfrac{1}{8}$

   $3\dfrac{11}{50}, 2\dfrac{1}{25}, \dfrac{3}{40}$

   $10\dfrac{7}{8}, \dfrac{7}{10000}, \dfrac{3}{2500}$

   $\dfrac{17}{50}, 11\dfrac{21}{2000}, 6\dfrac{1}{500}$

   $4\dfrac{23}{200}, \dfrac{7}{20}, \dfrac{191}{200}$

  2. 73.575, round to 73.6

Problem Set 3, pages 172–173 (continued)

D. 3. The same 6 digits repeat.

$$\frac{2}{7} = .\overline{285714}$$

$$\frac{3}{7} = .\overline{428571}$$

$$\frac{4}{7} = .\overline{571428}$$

$$\frac{5}{7} = .\overline{714285}$$

$$\frac{6}{7} = .\overline{857142}$$

4. $0.43

5. $7.88

6. 1.375 grams, $5\frac{1}{5}$ tablets

Chapter 4   Percent

Problem Set 1, pages 185–186

A. 40%, 10%, 95%
3%, 30%, 1.5%
60%, 1%, 120%
456%, 225%, 775%
.3%, 300%, 80%
550%, 400%, 604%
1000%, 33.5%

B. 20%, 75%, 70%
35%, 150%, 25%
10%, 50%, $37\frac{1}{2}\%$
60%, 175%, 220%
180%, 90%, $33\frac{1}{3}\%$
$216\frac{2}{3}\%$, $66\frac{2}{3}\%$, $68\frac{3}{4}\%$
$191\frac{2}{3}\%$, 330%

C. .07, .03, .56
.15, .01, .075
.90, 2.00, .003
.0007, .0025, 1.50
.015, .063, .005
.1225, 1.255, .667
.305, .085

D. $\frac{1}{10}$, $\frac{13}{20}$, $\frac{1}{2}$

$\frac{1}{5}$, $\frac{1}{4}$, $\frac{2}{25}$

$\frac{9}{10}$, $1\frac{7}{20}$, $\frac{3}{100}$

$\frac{3}{25}$, $\frac{1}{200}$, $\frac{3}{10000}$

$\frac{9}{200}$, $2\frac{1}{5}$, $\frac{3}{200}$

$\frac{1}{3}$, $\frac{31}{400}$, $\frac{13}{200}$

$\frac{1}{6}$, $\frac{1}{32}$

Problem Set 2, pages 201–202

A. 80%, 20%
    $9, 64%
    15, 107
    100%, 54
    21, 60
    $12\frac{1}{2}\%$, 150
    59, 80
    $66\frac{2}{3}\%$, 500%
    100, 52%
    500, $21.25

B. 225, 7755
    25%, 18
    $0.1974 or $0.20, 180%
    160, $2.72
    150%, $12\frac{1}{2}\%$
    $26\frac{2}{3}$, 4.5
    2%, $6\frac{2}{3}\%$
    17.5, 25%
    400%, 427
    22.1, 1600

C. 1. 88%   2. 16   3. 50.4
    4. no difference   5. 32.5% (rounded)   6. 5.5% (rounded)

Problem Set 3, pages 213–214

1. 29% (rounded)   2. $50,000   3. $33.70
4. $200
5. (a) $180, $4680
    (b) $0.08, $2.06
    (c) $0.03, $0.72
    (d) $0.43, $11.03
    (e) $0.13, $3.23
6. 16%
7. (a) $279.65
    (b) $28.76
    (c) $22.37
8. $107.25
9. $819 and $768
    The better loan is at Last National.
10. $122.50

ANSWERS FOR CHAPTER SELF-TESTS

If you answer any question incorrectly, turn to the page and frame indicated for review.

Chapter 1 Self-Test, page 75   Chapter 2 Self-Test, page 133

		page	frame			page	frame
1.	83	3	1	1.	$\dfrac{115}{16}$	79	1
2.	5221	3	1				
3.	164	3	1	2.	$3\dfrac{4}{11}$	79	1
4.	26	20	16				
5.	286	20	16	3.	$\dfrac{15}{40}$	79	1
6.	2028	20	16				
7.	3695	20	16	4.	$\dfrac{13}{17}$	79	1
8.	1548	29	24				
9.	48,348	29	24	5.	$\dfrac{31}{35}$	107	34
10.	55,622	29	24				
11.	650,206	29	24	6.	$1\dfrac{19}{60}$	107	34
12.	23	41	35				
13.	803 R6	41	35	7.	$6\dfrac{7}{24}$	107	34
14.	502	41	35				
15.	32	41	35	8.	$\dfrac{5}{12}$	107	34
16.	$2^3 \times 3 \times 17$	51	44				
17.	$2 \times 3 \times 11^2 \times 13$	51	44	9.	$1\dfrac{3}{20}$	107	34
18.	$2^3 \times 5^3$	51	44				
19.	$2^6 \times 3 \times 17$	51	44	10.	$3\dfrac{5}{12}$	107	34
20.	81	65	59				
21.	2000	65	59	11.	1	92	19
22.	7938	65	59				
23.	15,129	65	59	12.	$5\dfrac{1}{7}$	92	19
24.	15	65	59				
25.	18	65	59	13.	$1\dfrac{1}{9}$	98	26
				14.	$\dfrac{69}{154}$	98	26
				15.	$\dfrac{1}{4}$	92	19
				16.	$5\dfrac{4}{9}$	92	19
				17.	$58\dfrac{1}{2}$	92	19
				18.	$\dfrac{7}{16}$	120	51
				19.	$\dfrac{2}{5}$	120	51

Chapter 2 Self-Test, page 133

		page	frame
20.	$\frac{3}{4}$	120	51
21.	2	120	51
22.	$\frac{7}{25}$	120	51
23.	$12\frac{1}{4}$	120	51
24.	$\frac{9}{100}$	120	51
25.	72¢	120	51

Chapter 3 Self-Test, page 174

		page	frame
1.	19.245	137	1
2.	51.16	137	1
3.	41.151	137	1
4.	68.07	137	1
5.	0.47	137	1
6.	171.53	137	1
7.	75.15	137	1
8.	16.524	147	10
9.	168	147	10
10.	30.552	147	10
11.	1.90	153	16
12.	189.33	153	16
13.	0.794	153	16
14.	$\frac{14}{25}$	163	25
15.	$3\frac{31}{125}$	163	25
16.	$32\frac{13}{100}$	163	25
17.	0.4375	163	25
18.	3.625	163	25
19.	1.2	163	25
20.	0.6	163	25
21.	1.2	163	25
22.	14.95	163	25
23.	8.575	163	25
24.	7	163	25
25.	12.5	163	25

Chapter 4 Self-Test, page 215

		page	frame
1.	$316\frac{2}{3}\%$	177	1
2.	$41\frac{2}{3}\%$	177	1
3.	8%	179	4
4.	643%	179	4
5.	.02	182	7
6.	1.125	182	7
7.	$\frac{17}{25}$	183	9
8.	120	189	13
9.	7.56	189	13
10.	115.5	189	13
11.	8.71	189	13
12.	6.231	189	13
13.	35%	192	18
14.	150%	192	18
15.	225%	192	18
16.	42%	192	18
17.	$\frac{8}{27}\%$	192	18
18.	80	196	20
19.	.71 (rounded)	196	20
20.	75.5¢	203	24
21.	$62.00	203	24
22.	$59.20	203	24
23.	$5.37	205	29
24.	$72.50	210	34
25.	$1.97	208	32

ANSWERS TO FINAL EXAM, page 217

1. 349
2. 8975
3. 217
4. 1876
5. 1566
6. 285,324
7. 47
8. 457
9. $2 \times 3^3 \times 7$
10. $2^2 \times 5 \times 7 \times 11$
11. 16
12. $\dfrac{74}{9}$
13. $7\dfrac{5}{6}$
14. $\dfrac{5}{7}$
15. $1\dfrac{3}{20}$
16. $4\dfrac{19}{40}$
17. $9\dfrac{3}{4}$
18. $\dfrac{4}{15}$
19. $\dfrac{29}{14}$ or $2\dfrac{1}{14}$
20. $\dfrac{15}{24}$ or $\dfrac{5}{8}$
21. $2\dfrac{1}{7}$
22. $\dfrac{5}{12}$
23. $1\dfrac{11}{45}$
24. $21\dfrac{7}{9}$
25. $\dfrac{15}{68}$
26. 18.836
27. 164.195
28. 25.57
29. 7.73
30. 12.248

31. 40.42
32. 34.1220
33. .64
34. 1.88
35. $4\dfrac{69}{250}$
36. $.8\overline{3}$
37. .8
39. 1%
40. 1.28
41. 27
42. 7
43. 12.5
44. 5
45. 30%
46. 3742
47. 1,958,400 ft $\cong$ 371 miles
48. $37
49. 40¢
50. $3.99
51. $300
52. $6.75
53. $5.62
54. $\dfrac{17}{120}$
55. $2500

Number	Square Root	Number	Square Root	Number	Square Root	Number	Square Root
1	1.0000	51	7.1414	101	10.0499	151	12.2882
2	1.4142	52	7.2111	102	10.0995	152	12.3288
3	1.7321	53	7.2801	103	10.1489	153	12.3693
4	2.0000	54	7.3485	104	10.1980	154	12.4097
5	2.2361	55	7.4162	105	10.2470	155	12.4199
6	2.4495	56	7.4833	106	10.2956	156	12.4900
7	2.6458	57	7.5198	107	10.3441	157	12.5300
8	2.8284	58	7.6158	108	10.3923	158	12.5698
9	3.0000	59	7.6811	109	10.4403	159	12.6095
10	3.1623	60	7.7460	110	10.4881	160	12.6191
11	3.3166	61	7.8102	111	10.5357	161	12.6886
12	3.4640	62	7.8740	112	10.5830	162	12.7279
13	3.6056	63	7.9397	113	10.6301	163	12.7671
14	3.7417	64	8.0000	114	10.6771	164	12.8062
15	3.8730	65	8.0623	115	10.7238	165	12.8452
16	4.0000	66	8.1240	116	10.7703	166	12.8841
17	4.1231	67	8.1854	117	10.8167	167	12.9228
18	4.2426	68	8.2462	118	10.8628	168	12.9615
19	4.3589	69	8.3066	119	10.9087	169	13.0000
20	4.4721	70	8.3666	120	10.9545	170	13.0384
21	4.5826	71	8.4261	121	11.0000	171	13.0767
22	4.6904	72	8.4853	122	11.0454	172	13.1149
23	4.7958	73	8.5440	123	11.0905	173	13.1529
24	4.8990	74	8.6023	124	11.1355	174	13.1909
25	5.0000	75	8.6603	125	11.1803	175	13.2288
26	5.0990	76	8.7178	126	11.2250	176	13.2665
27	5.1962	77	8.7750	127	11.2694	177	13.3041
28	5.2915	78	8.8318	128	11.3137	178	13.3417
29	5.3852	79	8.8882	129	11.3578	179	13.3791
30	5.4772	80	8.9443	130	11.4018	180	13.4164
31	5.5678	81	9.0000	131	11.4455	181	13.4536
32	5.6569	82	9.0554	132	11.4891	182	13.4907
33	5.7446	83	9.1104	133	11.5326	183	13.5277
34	5.8310	84	9.1652	134	11.5758	184	13.5647
35	5.9161	85	9.2195	135	11.6190	185	13.6015
36	6.0000	86	9.2736	136	11.6619	186	13.6382
37	6.0828	87	9.3274	137	11.7047	187	13.6748
38	6.1644	88	9.3808	138	11.7473	188	13.7113
39	6.2450	89	9.4340	139	11.7898	189	13.7477
40	6.3246	90	9.4868	140	11.8322	190	13.7840
41	6.4031	91	9.5391	141	11.8743	191	13.8203
42	6.4807	92	9.5917	142	11.9164	192	13.8564
43	6.5574	93	9.6437	143	11.9583	193	13.8924
44	6.6332	94	9.6954	144	12.0000	194	13.9284
45	6.7082	95	9.7468	145	12.0416	195	13.9642
46	6.7823	96	9.7980	146	12.0830	196	14.0000
47	6.8537	97	9.8489	147	12.1244	197	14.0357
48	6.9282	98	9.8995	148	12.1655	198	14.0712
49	7.0000	99	9.9499	149	12.2066	199	14.1067
50	7.0711	100	10.0000	150	12.2474	200	14.1424

Study Cards

The following few pages contain useful information that you may want to remember. If you want to work on memorizing the multiplication table, the first few perfect squares, the first few primes, or other information, cut out the handy reminders from these pages and carry them with you in pocket or purse. Refer to them at every opportunity. Quiz yourself on them. Get a friend or tutor to quiz you on them. Read and recite the material until you have it firmly in your memory.

+	0	1	2	3	4	5	6	7	8	9
0	0	1	2	3	4	5	6	7	8	9
1	1	2	3	4	5	6	7	8	9	10
2	2	3	4	5	6	7	8	9	10	11
3	3	4	5	6	7	8	9	10	11	12
4	4	5	6	7	8	9	10	11	12	13
5	5	6	7	8	9	10	11	12	13	14
6	6	7	8	9	10	11	12	13	14	15
7	7	8	9	10	11	12	13	14	15	16
8	8	9	10	11	12	13	14	15	16	17
9	9	10	11	12	13	14	15	16	17	18

ADDITION TABLE

×	0	1	2	3	4	5	6	7	8	9	10
0	0	0	0	0	0	0	0	0	0	0	0
1	0	1	2	3	4	5	6	7	8	9	10
2	0	2	4	6	8	10	12	14	16	18	20
3	0	3	6	9	12	15	18	21	24	27	30
4	0	4	8	12	16	20	24	28	32	36	40
5	0	5	10	15	20	25	30	35	40	45	50
6	0	6	12	18	24	30	36	42	48	54	60
7	0	7	14	21	28	35	42	49	56	63	70
8	0	8	16	24	32	40	48	56	64	72	80
9	0	9	18	27	36	45	54	63	72	81	90

MULTIPLICATION TABLE

PERFECT SQUARES

$1^2 = 1$	$6^2 = 36$	$11^2 = 121$	$16^2 = 256$
$2^2 = 4$	$7^2 = 49$	$12^2 = 144$	$17^2 = 289$
$3^2 = 9$	$8^2 = 64$	$13^2 = 169$	$18^2 = 324$
$4^2 = 16$	$9^2 = 81$	$14^2 = 196$	$19^2 = 361$
$5^2 = 25$	$10^2 = 100$	$15^2 = 225$	$20^2 = 400$

THE PRIMES LESS THAN 100

2	3	5	7	11
13	17	19	23	29
31	37	41	43	47
53	59	61	67	71
73	79	83	89	97

if $\quad A \times B = C$

then $\quad B = C \div A$

and $\quad A = C \div B$

Percent	Decimal	Fraction	Percent	Decimal	Fraction
5%	.05	$\frac{1}{20}$	50%	.50	$\frac{1}{2}$
$6\frac{1}{4}\%$	.0625	$\frac{1}{16}$	60%	.60	$\frac{3}{5}$
$8\frac{1}{3}\%$	$.08\bar{3}$	$\frac{1}{12}$	$62\frac{1}{2}\%$	.625	$\frac{5}{8}$
10%	.10	$\frac{1}{10}$	$66\frac{2}{3}\%$	$.6\bar{6}$	$\frac{2}{3}$
$12\frac{1}{2}\%$	.125	$\frac{1}{8}$	70%	.70	$\frac{7}{10}$
$16\frac{2}{3}\%$	$.1\bar{6}$	$\frac{1}{6}$	75%	.75	$\frac{3}{4}$
20%	.20	$\frac{1}{5}$	80%	.80	$\frac{4}{5}$
25%	.25	$\frac{1}{4}$	$83\frac{1}{3}\%$	$.8\bar{3}$	$\frac{5}{6}$
30%	.30	$\frac{3}{10}$	$87\frac{1}{2}\%$	.875	$\frac{7}{8}$
$33\frac{1}{3}\%$	$.3\bar{3}$	$\frac{1}{3}$	90%	.90	$\frac{9}{10}$
$37\frac{1}{2}\%$	.375	$\frac{3}{8}$	100%	1.00	$\frac{10}{10}$
40%	.40	$\frac{2}{5}$			

SIGNAL WORDS　　　　　　　　TRANSLATE AS

is, is equal to, equals, the same as . . . . $=$

of, the product of, multiply, times, multiplied by . . . . . $\times$

add, in addition, plus, more, more than, sum, and,
increased by, added to . . . . $+$

subtract, subtract from, less, less than, difference,
diminished by, decreased by . . . . $-$

divide, divided by . . . . $\div$

twice, double, twice as much . . . . $2 \times$

half of, half . . . . $\frac{1}{2} \times$

THE MOST OFTEN USED SQUARE ROOTS

$$\sqrt{2} \cong 1.4142 \cong \frac{7}{5} \text{ or even closer } \frac{17}{12}$$

$$\sqrt{3} \cong 1.7321 \cong \frac{7}{4} \text{ or even closer } \frac{19}{11}$$

$$\sqrt{5} \cong 2.2361 \cong \frac{9}{4}$$

$$\sqrt{6} \cong 2.4495 \cong \frac{22}{9}$$

$$\sqrt{7} \cong 2.6457 \cong \frac{8}{3}$$

$$\sqrt{8} \cong 2.8284 \cong \frac{14}{5} \text{ or even closer } \frac{17}{6}$$

$$\sqrt{10} \cong 3.1623 \cong \frac{19}{6}$$

A MEMORY GIMMICK

If $A \times B = C$ then $A = \dfrac{C}{B}$ and $B = \dfrac{C}{A}$ for any numbers A, B, and C that are not zero.

Do you need a memory jogger? Try this one.

Examples:

$A \times B = C$ $=$ $3 \times 4 = 12$

$A = \dfrac{C}{B}$ $=$ $=$ $3 = \dfrac{12}{4}$

$B = \dfrac{C}{A}$ $=$ $=$ $4 = \dfrac{12}{3}$

Index